STUDY GUIDE
TO BECOME A
CISCO FIREWALL ENGINEER

BY

ISAAC DANSO-GYAN

www.networkprofessional.net training@networkprofessional.net

ISBN

Hardcover: 978-1-969844-61-4

Paperback: 978-1-969844-60-7

Table of Contents

Introduction

In today's complex cybersecurity landscape, **firewalls remain the first and most critical line of defense**. Cisco, a global leader in networking and security solutions, has continuously evolved its firewall technologies over more than two decades — from the early **PIX firewalls** and **ASA platforms** to the modern **Firepower Threat Defense (FTD)** and **Secure Firewall** series.

Cisco's firewall roadmap reflects a **shift from hardware-focused packet filtering** to **intelligent, policy-driven security systems** that integrate with cloud environments, automation frameworks, and AI-powered threat intelligence.

The **Cisco ASA (Adaptive Security Appliance)**, first introduced in the mid-2000s, quickly became the industry standard for enterprise firewalls. It combined powerful packet inspection, VPN support, and NAT control with reliability and performance. For many years, ASA devices formed the backbone of secure network perimeters across corporations, data centers, and service providers.

As security threats evolved, Cisco expanded its capabilities with **Firepower Threat Defense (FTD)** — a next-generation firewall platform that unified ASA's proven foundation with **advanced intrusion prevention (IPS), application visibility, URL filtering, and malware protection**. Managed through the **Firepower Management Center (FMC)** or **FDM (Firepower Device Manager)**, FTD introduced automation, analytics, and deep threat intelligence, marking a major leap forward in Cisco's security roadmap.

This book bridges the gap between the **traditional ASA architecture** and the **modern FTD ecosystem**. It helps you:

- Understand the **evolution of Cisco firewall technologies**
- Configure both ASA and FTD in real-world environments
- Learn the **core principles** that apply across all Cisco security platforms
- Prepare for **certifications or production deployments** with clarity and confidence

By focusing on what matters most — **concepts, configuration, and context** — this guide eliminates unnecessary complexity and gives you a fast path to becoming proficient with Cisco firewalls

Routing

Routing protocols are used to determine how data is transmitted between networks. They can be categorized as link-state, distance vector, or exterior gateway protocols.

In simple terms, routing protocols determine how routers communicate with each other to share information about network topology. They allow routers to dynamically adapt to changes, such as link failures or network congestion, ensuring reliable and efficient data delivery. Understanding these protocols is crucial for network engineers and administrators tasked with maintaining seamless connectivity.

Routing Protocols

These are protocols which help Routing Protocols to carry their information from one router to another example: **Static routing** and **dynamic routing** such as OSPF, EIGRP, RIP, IS-IS, and BGP, to figure out what paths traffic should take

Network Routing Recap Project

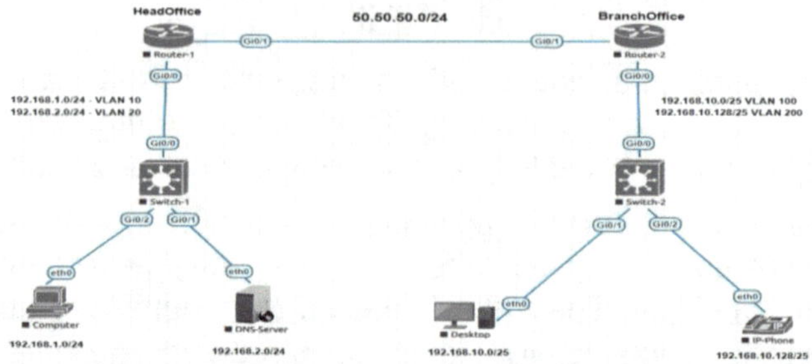

Global Configuration

1. Configure the hostname base on the Network Diagram.
2. Disable the DNS lookup feature.
3. Assign IKE as the Secret password.
4. Direct the Cisco IOS to encrypt any passwords stored in clear text.

Console Port

5. Configure the console port on all devices to log input synchronously.
6. Set password to NPTC.
7. Configure the idling timeout to 30 minutes.

VTY Ports

8. Allow 5 concurrent sessions of remote access.
9. Configure the vty ports to log input synchronously.
10. Set password to V
11. Configure idling timeout to 30 minutes and 10 seconds.
12. Save config.

Verify the above steps using the proper Show command

13. Assigning IP Addresses and Port Description
14. Configure the Branch Office to act as a DHCP Server and exclude 10 IP addresses from the VLAN 200 Scope

    VLAN and Trunk

15. Configure VLANs 10 and 20 on Switch 1 and name them as Desktop and Servers, respectively
16. Configure VLANs 100 and 200 on Switch 2 and name them as Desktop and VOIP, respectively
17. Configure the switch virtual interface (SVI) using the respective VLAN on the Switch
18. Configure a Switch Default Gateway
19. Configure the Trunk Port base on the topology
20. Configure the Access port base on the topology
21. Disable all the ports on the switches that are not connected.

First Hop Redundancy Protocol" (FHRP)

A "First Hop Redundancy Protocol" (FHRP) is a networking protocol designed to provide redundancy for the default gateway on a network, ensuring seamless connectivity even if the primary router fails by allowing multiple routers to share a virtual IP address and act as a backup for one another, with only one active at a time; essentially, it protects against single points of failure at the first hop of a network connection.

Key Points about FHRP:

Purpose:

To maintain network connectivity by automatically switching to a backup router if the primary gateway becomes unavailable.

Virtual IP Address:

All routers in an FHRP group share a single "virtual IP address," which is used as the default gateway by devices on the network.

Active and Standby Routers:

Within an FHRP group, only one router is designated as "active" and handles traffic, while the others remain in "standby" mode, ready to take over if needed.

Failover Mechanism:

When the active router fails, the FHRP protocol detects the issue and automatically elects a new active router from the standby group, ensuring minimal disruption to network traffic.

TYPES of FHRP

Hot Standby Router Protocol (HSRP): A Cisco proprietary protocol considered the most widely used FHRP.

Virtual Router Redundancy Protocol (VRRP): An industry standard FHRP with similar functionality to HSRP.

Gateway Load Balancing Protocol (GLBP): Allows for load balancing across multiple active routers, not just failover.

Redundancy Protocol Project

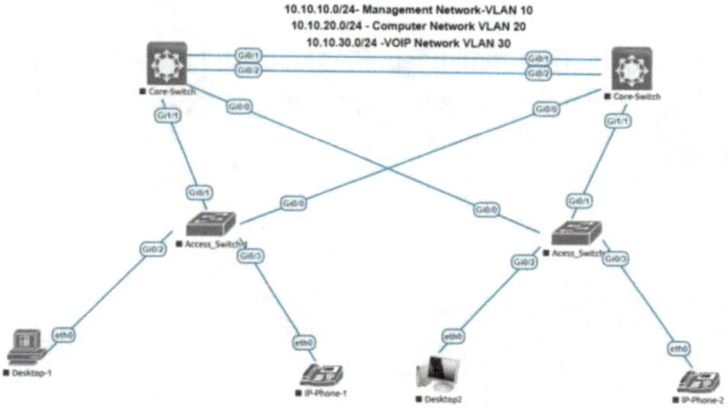

Global Configuration

1. Configure the hostname base on the topology.
2. Disable the DNS lookup feature.
3. Assign Cisco as the Secret password.

Console Port

4. Configure the console port on all devices to log input synchronously.
5. Set password to Cisc0
6. Configure the idling timeout to 30 minutes.

VTY Ports

7. Allow 5 concurrent sessions of remote access.
8. Configure the VTY ports to log input synchronously.
9. Set password to Cisc0
10. Configure idling timeout to 60 minutes.

VLAN Configuration

11. **Configure the below VLANs as follows and allow them to propagate to the access switch.**

VLAN 10 - Management VLAN - **10.10.10.0/24**

VLAN 20 - Computer VLAN – **10.10.20.0/24 and create as Multicast**

VLAN 30 - VOIP VLAN **– 10.10.30.0/24**

Trunk Configuration

12. **Configure the trunk port base on the topology.**

VLAN Trunking Protocol (VTP) Configuration

13. Configure the VTP mode of the access switch as Client
14. Configure the VTP domain name and password on both Core switches as follows

Domain Name – NPTC

Password- secret

EtherChannel's

15. Configure EtherChannel using PAgP with ON mode on Core_SW1 and ON mode on Core_SW2.

Redundancy Configuration

16. Configure your network using HSRP, the above ether channel topology.

IP Routing

17. Enable IP routing on the two Core switches.

Access Port

18. Configure all access ports for auto settings.

Introduction of a Firewall in a Computer Network

A firewall **is a** network security device **or software that monitors and controls** incoming and outgoing network traffic **based on predefined** security rules**. It acts as a** barrier between a trusted internal network and untrusted external networks**, such as the internet.** The primary purpose of a firewall is to separate a secure area (Higher Security Zone/Inside Network) from a less secure area (Lower Security Zone/Outside Network) and to control communication between the two. Firewall also controls inbound and outbound communications across devices.

They work by examining incoming and outgoing data packets and comparing them against a set of rules. Based on these rules, the firewall can either allow or block the traffic.

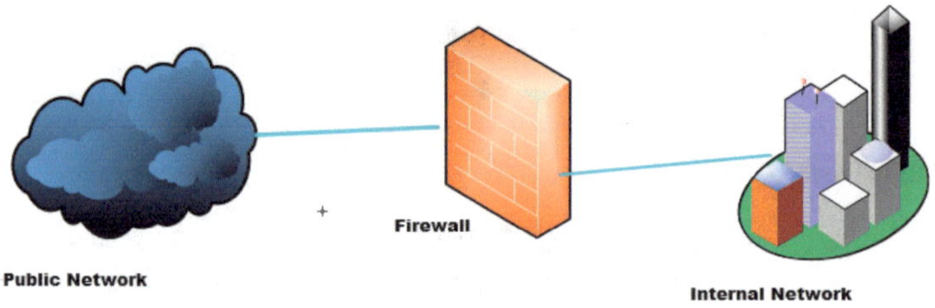

In large corporate network environments, you can also place a network firewall within your internal LAN to provide segmentation of private LAN IP subnets (e.g., isolating the server's LAN from the users' LAN, for example).

Throughout the years, firewalls have started to include IDS and IPS functions, as well as Anti-X and Web content filtering services.

Firewall Products

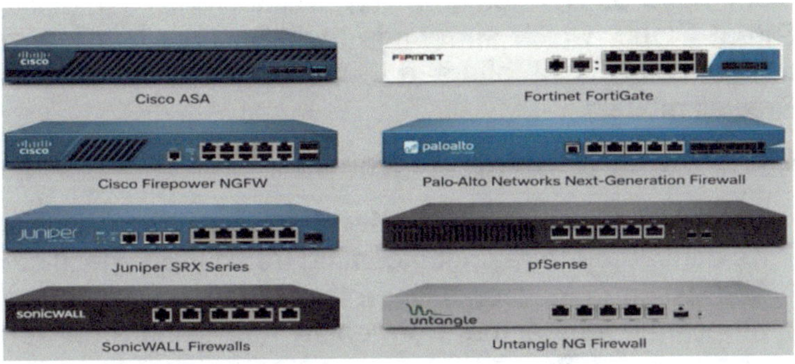

Competitors to Cisco ASA

Cisco ASA with Firepower services is a premium security product for Enterprise Networks, and according to Gartner.com and Spiceworks.com, there are only three direct competitors to these Cisco products. They are Palo Alto, Fortinet, and Check Point.

Palo Alto

Palo Alto next-generation firewalls provide similar features to Cisco ASA firewalls through their PAN-OS operating system. The Palo Alto firewalls and firewall clusters can be managed by their Firewall management system, known as Panorama.

Fortinet

Fortinet offers a wide range of firewall models designed for every network size, from entry-level to cloud data centers. These firewalls run the FortiGate operating system. Fortinet is one of the fastest-growing security firms worldwide, and it manufactures a wide range of security products, including firewalls, antivirus, email security, SIEM, and Wi-Fi solutions.

Checkpoint

Checkpoint has taken a unified approach to network security through a suite of products that includes Next-Generation Firewalls, known as the Infinity architecture.

This architecture is composed of five sections: Quantum, Cloud Guard, Harmony, Infinity Vision, and the Infinity Threat Cloud, which surrounds their Security Intelligence center. Checkpoint offers a wide range of 15 different Firewall models.

Advantages of Using A Firewall

1. **Protection from unauthorized access:** Firewalls can be configured to restrict incoming traffic from specific IP addresses or networks, thereby preventing hackers or other malicious actors from easily accessing a network or system. Protection from unwanted access.
2. **Prevention of malware and other threats:** Malware and other threat prevention: Firewalls can be set up to block traffic associated with known malware or other security concerns, thereby assisting in defense against these types of attacks.
3. **Control of network access:** By limiting access to specified individuals or groups for particular servers or applications, firewalls can be used to restrict access to particular network resources or services.
4. **Monitoring of network activity:** Firewalls can be configured to record and track all network activity. This information is essential for identifying and investigating security issues and other forms of suspicious behavior.
5. **Regulation compliance:** Many industries are bound by rules that demand the use of firewalls or other security measures. Organizations can comply with these rules and prevent any fines or penalties by using a firewall.
6. **Network segmentation:** By using firewalls to split up a bigger network into smaller subnets, the attack surface is reduced and the security level is raised.

Types of Firewall: Base, How to Deploy

1. **Dedicated hardware appliances** are generally used in data centers.
2. **Software on a machine** used by home users. e.g., Windows Firewall

3. **Managed firewall services** offer various options, including premises-based, network-based, or cloud-based services **(Firewall as a Service)**. In this case, the firewall manufacturer or service provider manages the network and is responsible for firewall administration, log monitoring, and other related tasks.

Types of Firewalls Based On Method of Operation

1. **Packet Filtering/Stateless Firewall:** As the name suggests, the user can either allow or drop packets based on source and destination IP, IP protocol ID, etc., from entering the internal network. This type of filtering works at the network transport layer.
2. **Proxy Firewall:** It offers more security than other types of filtering. In proxy filtering, the client connects with a proxy instead of a target system and initiates a new connection. This makes it harder for an attacker to discover the network, as they are not getting a response from the target system.
3. **Stateful Inspection Firewall:** In this type of inspection, systems maintain a state table (maintains active connections), analyze incoming and outgoing packets, and drop them accordingly.
4. **Application Layer Firewalls:** These firewalls can examine application layer (of the OSI model) information, like an HTTP request. If it finds a suspicious application that can be responsible for harming our network or that is not safe for our network, then it gets blocked right away.
5. **Next-generation Firewalls:** These firewalls are called intelligent firewalls. These firewalls can perform all the tasks that are performed by the other types of firewalls that we learned previously, but on top of that, they include additional features like application awareness and control, integrated intrusion prevention, and cloud-delivered threat intelligence.

Firewall Modes

Cisco ASA can be used in 2 modes, which are Routed Mode and Transparent Mode.

Routed Firewall Mode

In routed mode (default mode), the ASA is considered to be a router in the network. Routed mode supports many interfaces. Each interface is on a different subnet. The ASA acts as a router between connected networks, and each interface requires an IP address on a different subnet.

Transparent Firewall Mode.

ASA in Transparent firewall mode works as a layer 2 switch/bridge while still providing firewall benefits (intrusion prevention, packet inspection, etc).

Only the management interface can receive an IP address when the ASA is working in Transparent Mode. The ASA connects the same network between its interfaces. Because the firewall is not a routed hop, you can easily introduce a transparent firewall into an existing network without having to make any changes in the network.

<u>**Initial Configuration**</u>

Asa(config)# firewall transparent

Characteristics of Transparent Mode

- Transparent firewall mode supports only two interfaces (inside and outside)
- The firewall bridges packets from one VLAN to the other instead of routing them.
- MAC lookups are performed instead of routing table lookups.
- Can run in a single firewall context or in multiple firewall contexts.

- A management IP address is required on the ASA.
- The management IP address must be in the same subnet as the connected network.
- Each interface of the ASA must be a different VLAN interface.
- Even though the appliance acts as a Layer 2 bridge, Layer 3 traffic cannot pass through the security appliance from a lower security level to a higher security level interface.
- The firewall can allow any traffic through by using normal extended Access Control Lists (ACLs).

Benefits of Using A Firewall in Transparent Mode –

- No change required on existing IP addressing
- Protocols such as HSRP, VRRP, and GLBP can pass.
- Multicast streams can traverse
- Non-IP traffic can be allowed (IPX, MPLS, BPDUs, etc.)
- Routing protocols can establish adjacencies through the firewall

Cisco Firewall Models

Cisco PIX 500 Series (legacy) - The **Cisco PIX 500 Series** was a line of **dedicated hardware firewalls** introduced by Cisco Systems in the late 1990s and widely deployed throughout the early 2000s. These appliances are now considered **legacy** and have been **officially retired** and replaced by more advanced platforms such as the **Cisco ASA Series** and later the **Cisco Firepower** line

Models: 501 (user-based), 506, 515,510, 525, 535

Security Licensing Services: Stateful Firewall, IPsec VPN (Client, Site), Application Inspection (Using Fixups)

Cisco ASA 5500 Series

The **Cisco ASA 5500 Series** (Adaptive Security Appliance) is a family of hardware-based security devices developed by Cisco Systems. These appliances are widely used for **network security**, **VPN (Virtual Private Network)** services, **firewall protection**, **intrusion prevention**, and **remote access**. The ASA 5500 Series was a major part of Cisco's network security portfolio, though many models are now **end-of-life (EOL)** and have been replaced by the **Cisco ASA 5500-X Series** and **Cisco Firepower NGFW**

Common Models:

- **ASA 5505**: Entry-level, ideal for small offices.
- **ASA 5510 / 5520 / 5540 / 5550**: Mid-range models suitable for medium to large networks.
- **ASA 5580**: High-end model for data centers and enterprise environments.

Limitations:

- **End-of-Life**: Most ASA 5500 models are no longer supported. Cisco encourages migration to newer platforms.
- **Performance**: Limited compared to modern firewall appliances, especially with deep packet inspection and threat intelligence features.

Migration Path:

Cisco recommends upgrading to:

- **ASA 5500-X Series** (with Firepower services)
- **Cisco Firepower 1000, 2100, 4100 Series NGFWs** (Next-Gen Firewalls)

Security Licensing Services :(Verification command = Show license features)

Stateful Firewall • IPSec VPN (Client, Site) • SSL VPN (Client) • Application Inspection (using MQC- same as QOS in IOS) • Virtualization (using Contexts) • Modes: Transparent, Routed • Content Security (Anti-Spam, Anti-Phishing, Anti-Virus, Anti-Spyware, File Blocking, URL Filtering)• IPS

Cisco ASA Components

Ethernet ports

Management port

Console port

Flash Memory

Security Expansion

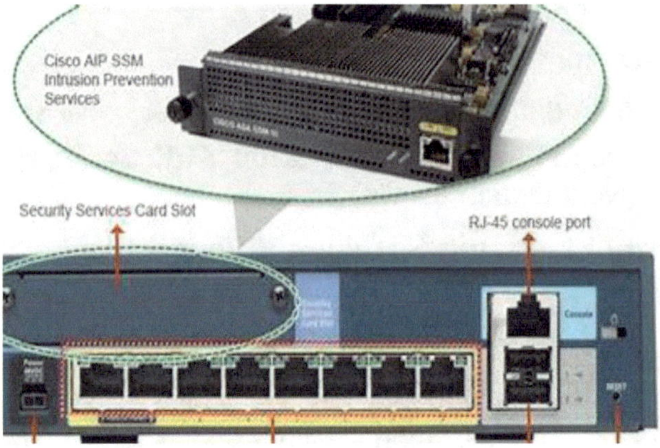

Security Service Card

A **Security Service Card** (also known as a **security module**, **service module**, or **security blade**) is a **hardware add-on component** used in network security appliances like firewalls and routers. It provides **specialized security processing capabilities**, enabling **advanced threat detection, content filtering, VPN acceleration**, and other services **without overloading the main system CPU**

- SSC (Security Services Card)
  - ASA 5505
  - Services: IPS

- SSM (Security Services Module)
  - ASA 5510, 5520, 5540
  - Services: IPS, Content Security

ASA Security Models (for High-end ASA Firewall)

- SSP (Security Services Processors)
  - ASA 5580, 5585
  - IPS services

Service Modules (Cisco Catalyst 6500)

FWSM-Firewall service models- This model normally doesn't have VPN features. The WEB VPN services module supports up to 32,000 SSL VPN users, and up to 4 modules can be used in a single chassis.

ACE- Application Control Engine Module for Cisco 6500 Series and 7600 Series routers (ACE supports translation and load balancing).

CSM-Content Switching Module

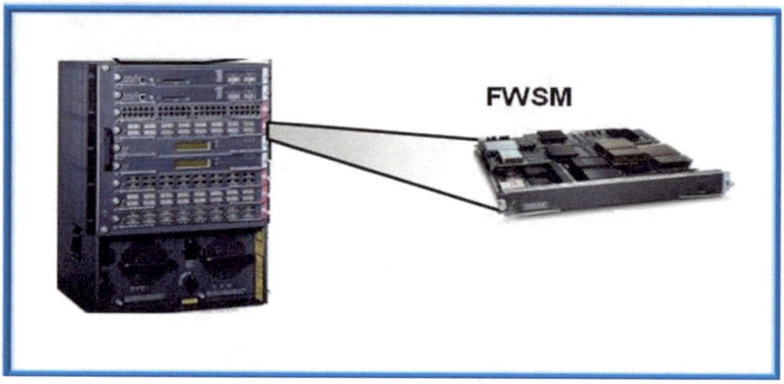

Performance Metric Consideration

- Firewall and VPN Services
  - Using one or the other not both!

High-end Network Security Appliances | Mid-range Network Security Appliances

Cisco ASA 5500 Series Model/License	Cisco ASA 5505 Base / Security Plus	Cisco ASA 5510 Base / Security Plus	Cisco ASA 5520	Cisco ASA 5540	Cisco ASA 5550
Product Image (click to enlarge)					
Network Location	Small Business, Branch Office, Enterprise Teleworker	Internet Edge	Internet Edge	Internet Edge	Internet Edge, Campus
Performance Summary					
Maximum Firewall throughput (Mbps)	150 Mbps	300 Mbps	450 Mbps	650 Mbps	1 Gbps (real-world HTTP), 1.2 Gbps
Maximum Firewall Connections	10,000 / 25,000	50,000 / 130,000	280,000	400,000	650,000
Maximum Firewall Connections/Second	4000	9000	12,000	25,000	36,000
Packets Per Second (64 byte)	85,000	190,000	320,000	500,000	600,000
Maximum 3DES/AES VPN Throughput	100 Mbps	170 Mbps	225 Mbps	325 Mbps	425 Mbps

Cisco ASA Firewall Security Levels

The Cisco ASA Firewall uses so-called "security levels" that indicate how trusted an interface is compared to another interface. The higher the security level, the more trusted the interface is. Each interface on the ASA is a security zone, so by using these security levels, we have different trust levels for our security zones.

An interface with a high-security level can access an interface with a low-security level, but the other way around is not possible unless we configure an access list that permits this traffic.

Rules

In short, this is how the security levels work:

- Traffic from a **higher security level to a lower security level is allowed**. For example, traffic from the inside is allowed to reach the outside. Of course, it's possible to restrict this with access lists.
- Traffic from a **lower security level to a higher security level is not allowed**. This could be traffic from the outside headed towards the inside. You can also change this with an access list. This might be

useful if you have servers in the DMZ that you want to reach from the outside.

- Traffic between interfaces with the **same security level is not allowed**. For example, if you have an interface called "DMZ1" with security level 50 and another one called "DMZ2" with the same security level, then traffic between the two will be dropped. You can change this behavior with the global **same-security-traffic permit inter-interface** command.
- Our LAN is our trusted network, which has a high security level. The WAN is untrusted, so it will have a low security level. This means that traffic from our LAN > WAN will be permitted. Traffic from the WAN to our LAN will be denied. Since the firewall is stateful, it keeps track of outgoing connections and will permit the return traffic from our LAN.
- If you want to make an exception and permit traffic from the WAN to the LAN, then this can be accomplished with an access list.
- Most companies will have one or more servers that should be reachable from the Internet. Perhaps a mail or web server. Instead of placing these on the INSIDE, we use a third zone called the **DMZ (Demilitarized Zone)**. Take a look at the picture below:

What is a Security Level?

A **security level** is a **numerical value** between **0 and 100**.

It represents the **relative trust** of an interface:

- **100** = Most trusted (usually **inside** network)
- **0** = Least trusted (usually **outside** or **internet**)
- Any value between 1–99 is typically assigned to **DMZs**, guest networks, or specialized zones.

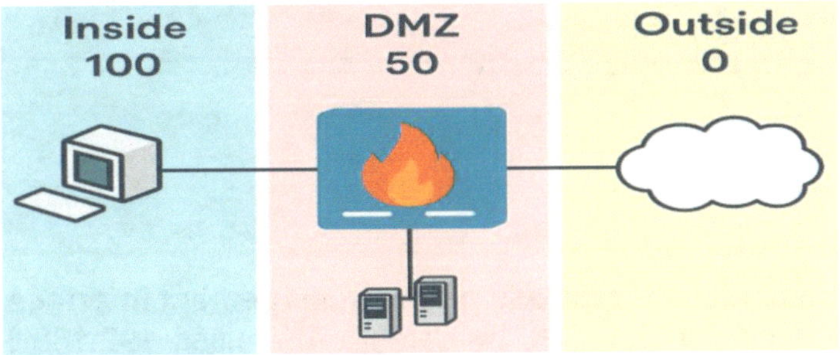

Default Behavior Based on Security Levels

From Interface	To Interface	Allowed by Default?
Higher security level → Lower level	✅ Yes	
Lower security level → Higher level	❌ No (ACL needed)	
Same security level → Same level	❌ No (unless allowed via special command)	

Cisco ASA ASDM Initial Configuration

Cisco's ASDM (Adaptive Security Device Manager) is the GUI that Cisco offers to configure and monitor your Cisco ASA firewall.

1. Task: Apply the Initial Setup to Get the Asdm Working for Management

ASDM Initial Setup Lab Project_1

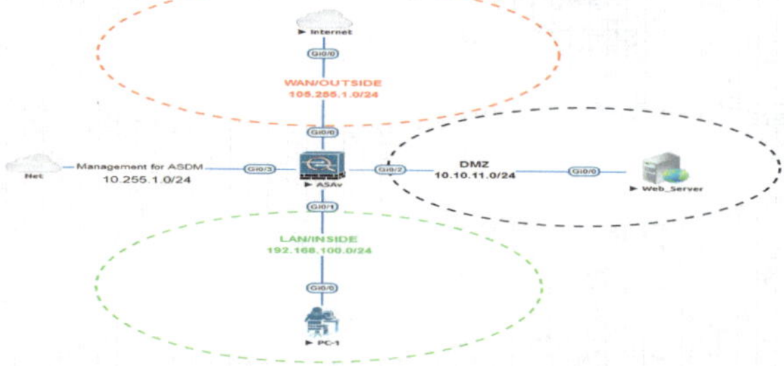

1. **Assign an IP address on the management interface to be used for the GUI,** by default, ASA uses 192.168.1.1, which can be changed

Boston-ASAFW(config)# interface GigabitEthernet0/3

Boston-ASAFW(config-if)#nameif OUTSIDE

Boston-ASAFW(config-if)#security-level 0

Boston-ASAFW (config-if)#ip address 10.255.1.101 255.255.255.0

2. **ASDM requires HTTP, and it's disabled by default**. Let's enable it:

Boston-ASAFW (config)#http server enable

3. Instead of giving everyone access to the HTTP server, we will **specify which network and interface are permitted to use the HTTP server:**

ASA-FW (config) # http 10.255.1.0 255.255.255.0 Management

4. Create a user account to be used by the ASDM

ASA-FW (config)#Username admin pass cisco

That's all we have to do on the ASA. Now you can open a web browser on your computer. Open the following URL:

The ASA uses a self-signed certificate so which is why you see this error above. Just click on Continue to this website, and you will see the following screen:

https://10.255.1.101

5. You now have two options…you can run ASDM directly from the ASA's flash memory, or you can install it on your computer first. This lab has it installed already

Enter the IP address of the ASA and the username/password that we created earlier. Click on OK, and you will see this:

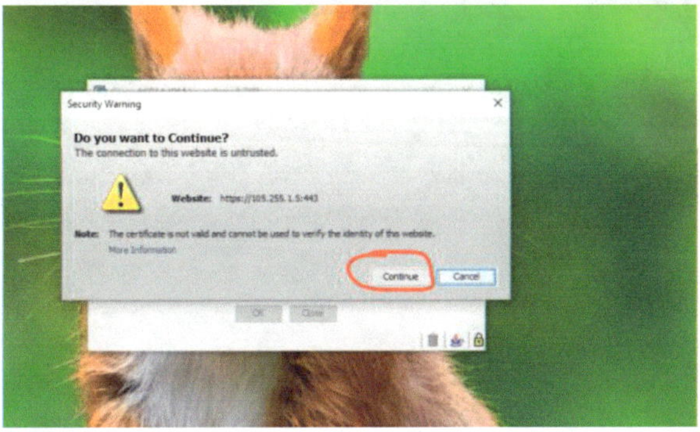

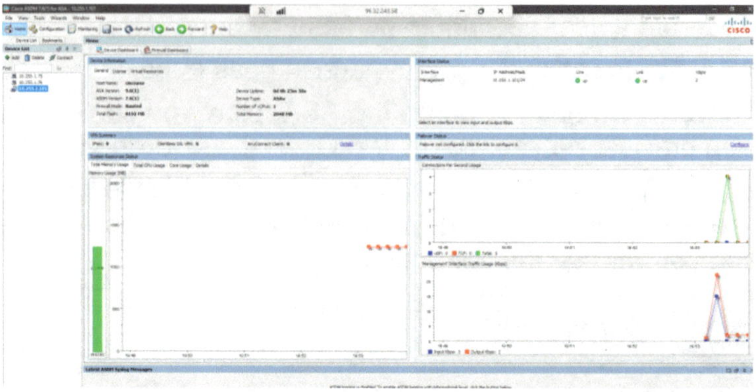

Using ASDM For Basic Configuration And AAA Access

1. Configure the hostname, domain name, and the enable password

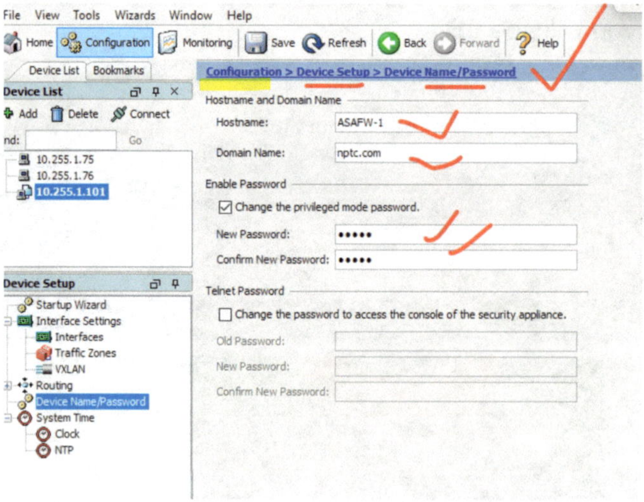

2. Configure the clock time, date, and time zone.

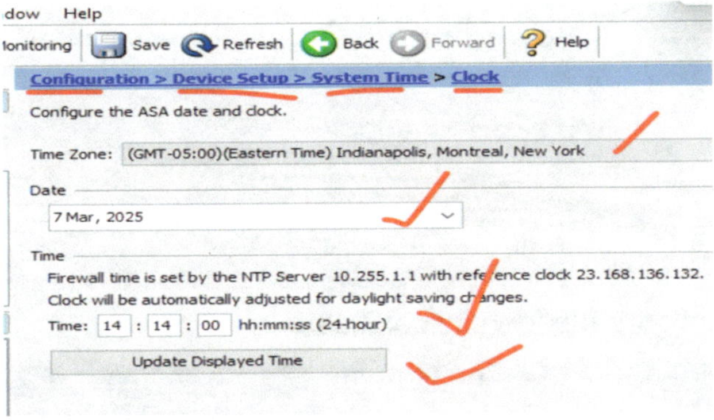

3. Configure the NTP server.

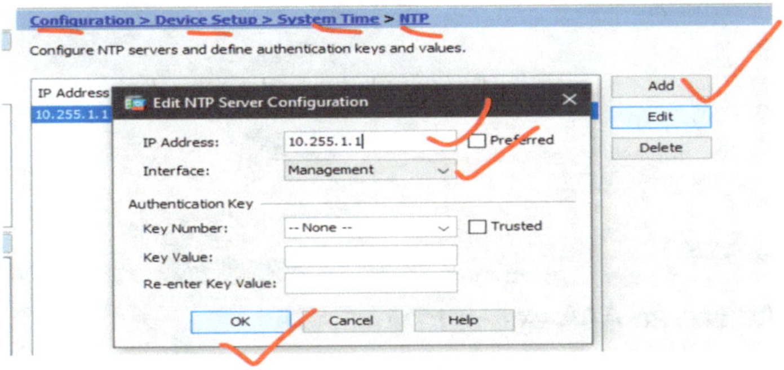

4. Create a user account.

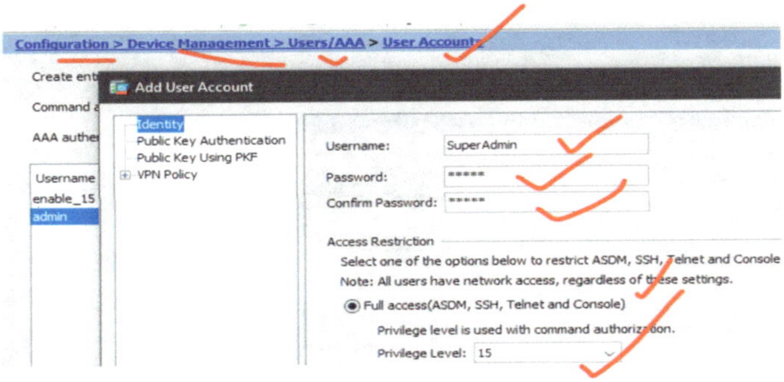

5. Giving AAA access to an Account.

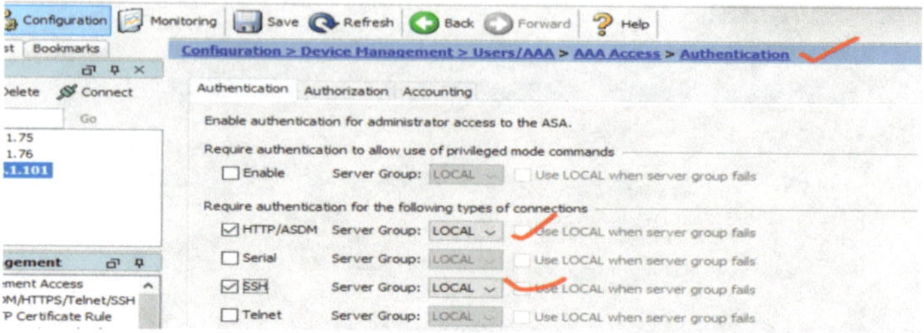

6. Configure the identity certificate.

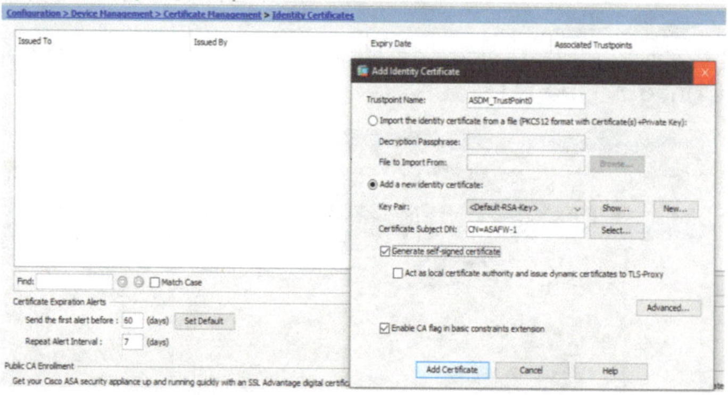

7. Adding an AAA Server to your ASA

The first step is to create the server group and add the server to the group.

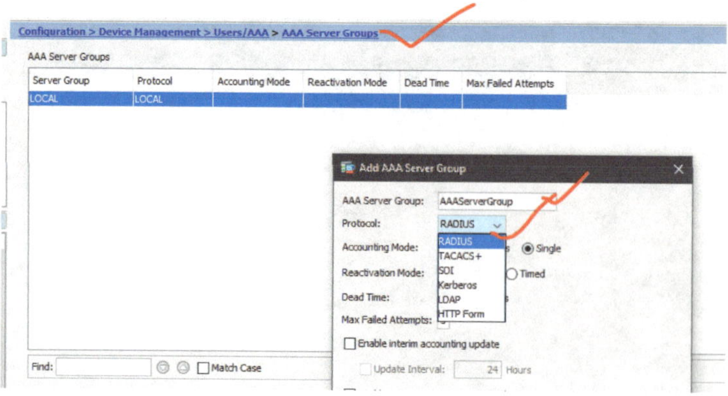

Now we can add a server to the group.

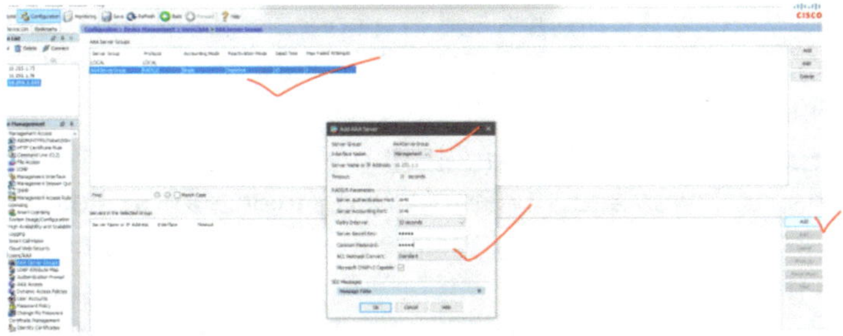

8. Enable access to the box with the new group.

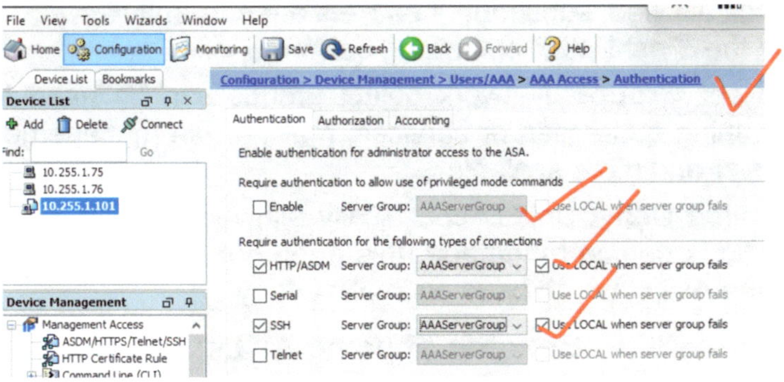

Cisco ASA Access Control-List (Firewall Policy)

A **Cisco ASA Access Control List (ACL)** is a set of rules that defines whether to permit or deny network traffic based on criteria like source, destination, protocol, and port. ACLs are fundamental to network segmentation and security on the Cisco ASA firewall, governing how and which traffic is allowed to flow between network zones—such as from a lower-security "outside" interface to a higher-security "inside" one

Key Features and Behavior

Direction: ACLs are applied in a specific direction (inbound or outbound) on an interface, but only one ACL may be assigned per interface, per direction.

Security Levels: By default, the ASA allows traffic from higher to lower security levels (e.g., inside to outside), but not from lower to higher (outside to inside) unless permitted by an ACL.

Processing: When a new connection attempt is detected, the ASA inspects the session's first packet against the ACL. Once allowed, subsequent packets are efficiently forwarded via the "Fast Path," optimizing performance

Object Groups: Network object groups can be used for grouping addresses or services, making rule management easier and more scalable

Implicit Deny: Unless specifically allowed, all unmatched traffic is denied by default (this does not apply to control-plane ACLs, which have no implicit deny

Benefits of using ACLs:

- **Enhanced Security:** ACLs help to prevent unauthorized access to network resources.
- **Traffic Control:** ACLs allow administrators to control the flow of network traffic, improving network performance and security.
- **Granular Control:** ACLs provide granular control over network traffic, allowing administrators to fine-tune security policies.

Types ACL

ACLs come in 2 main types used in ASAs: Standard, Extended. Each ACL type has a different application, depending on where it's deployed.

Standard. A standard ACL is designed to protect a network using only the destination address. These are typically used in simple deployments, and are used by only a few protocols like VPN filters and route maps (though route maps can also use extended ACLs, so it's rarely used in this case either). Standard ACLs do not provide robust security.

Extended. Building on a standard ACL, using extended ACLs means you can also allow or block source addresses in addition to the destination. Extended ACLs can also be applied to traffic based on a variety of protocols: IP, ICMP, TCP, and UDP, as well as service policies, AAA rules, WCCP, Botnet Traffic Filter, and VPN group, and DAP policies. Among the most common ACLs you will encounter.

Firewall rules

Firewall rules are often based on port numbers, specifying which ports can be accessed from specific IP addresses or networks.

By controlling which ports are open, firewalls can prevent unauthorized access to services and applications, protecting the network from malicious traffic.

Port Numbers

In the context of firewalls, port numbers are crucial for identifying specific services or applications on a network, allowing firewalls to control network traffic based on which ports are being used.

Port number is not a physical connection but a logical connection that is used by **programs** and **services** to exchange information.

It basically determines which programs or services on a computer or server are going to be used.

Port numbers range from 0 – 65535

The port number is always associated with an IP address to exchange data. The IP address determines the location of the server or computer, and the port number determines which application or program on the server it wants to use

Systems or Well-Known Ports

Ranges from 0-1023

Commonly used TCP PORT

80,443 – Web pages (HTTP, HTTPS)

21- FTP (File Transfer Protocol)

25- Email (SMTP)

53 – DNS communication

TCP: Real-life applications

Web browsing, Email, FTP, Remote Desktop

UDP Real Life applications

Online Gaming, Voice over IP, Streaming Video, DNS

Register Ports or User Port

Ranges from 1024 – 49151; these are ports that can be registered by companies and developers for particular services.

1102 – Adobe Server

1433- Microsoft SQL Server

1416- Novell

1527 – Oracle

Dynamic Port Numbers

Ranges 49152 – 65535, these are ports that your computer assigns temporarily to itself during a session. They are client-side ports that are free to use.

To check active connections on your computer, type netstat –n / netstat-an (if your computer is acting as a server)

ACL -Lab Project _1

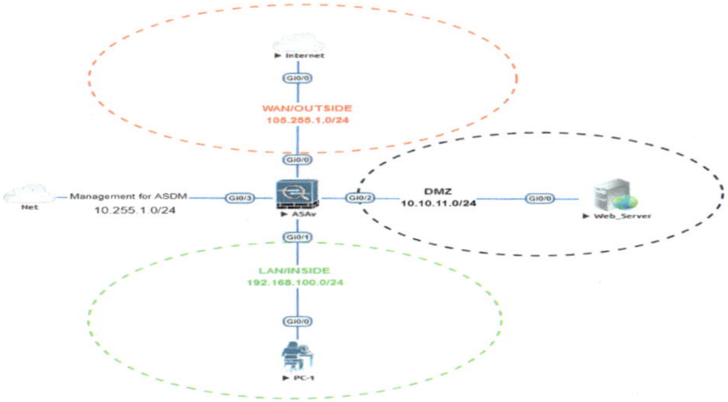

1. Project Task 1: **Configure a firewall policy that will permit any source outside (any IP address) access the web server 10.10.11.2 on TCP port 80 (WWW) using ASDM**

1. **Assign an IP address to the LAN, DMZ, and the outside interface base on the topology using ASDM**

Configuration>Device setup>interfaces

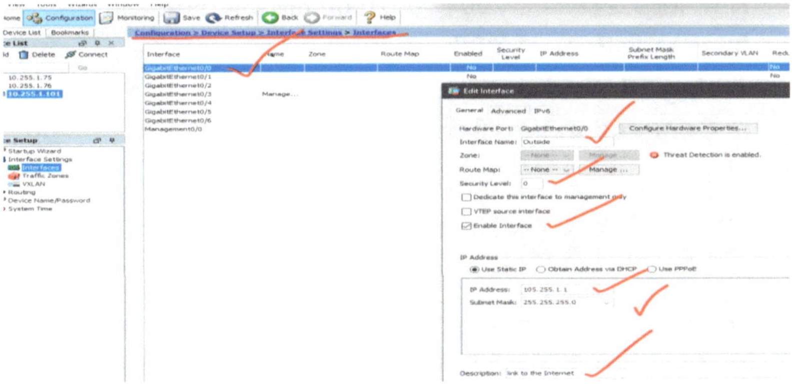

2. **Repeat the same process for the remaining interfaces**

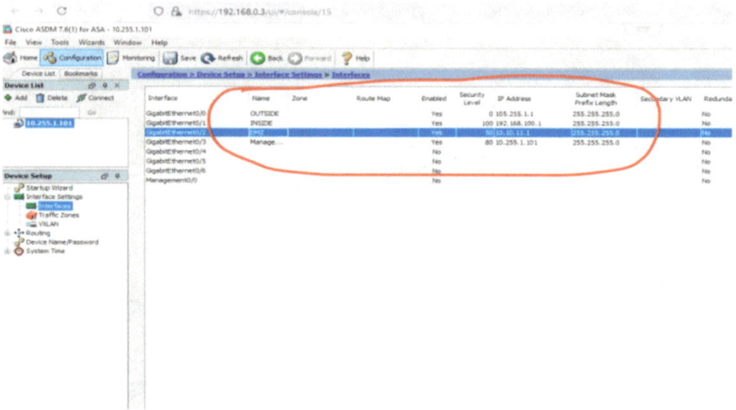

NB: But users from the LAN with a higher security level can communicate on the service port 80

PC-1#10.10.11.2 80

Trying 10.10.11.2, 80 ... Open

Now let's test if port 80 is open on the internet

Internet#10.10.11.2 80

Trying 10.10.11.2, 80 ...

% Connection timed out; remote host not responding

This traffic is denied by default because it is coming from a low Security Zone to a high Security Zone. Let's create an access-list that allows HTTP traffic. We'll create something so that users on the internet are allowed to connect to the web server on port 80. All other traffic will be denied.

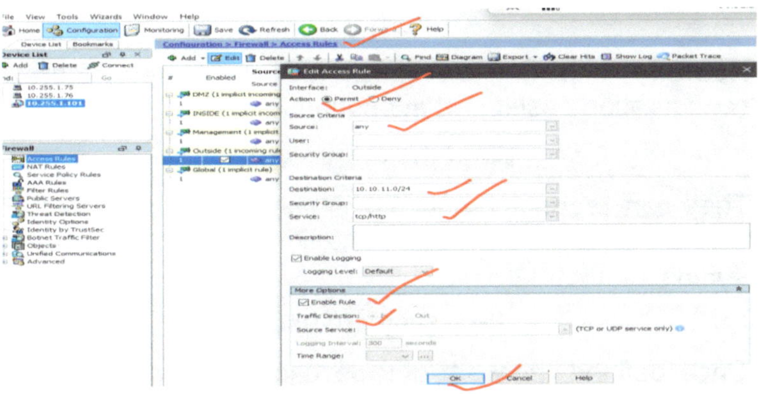

Let's do the testing on the internet to see if the port can open.

Internet#

Internet#

Internet#10.10.11.2 80

Trying 10.10.11.2, 80 ... Open

Network object in Cisco ASA firewall

In a Cisco ASA firewall, network objects are reusable components that represent IP addresses, subnets, or FQDNs, simplifying configuration and maintenance by allowing you to reference them in multiple rules and policies instead of repeating the same values.

Imagine you have to manage a Cisco ASA firewall that has hundreds of hosts and dozens of servers behind it, and for each of these devices, you require access-list rules that permit or deny traffic.

With so many devices, you will have a LOT of access-list statements, and it might become an administrative nightmare to read, understand, and update the access-list.

To make our lives a bit easier, Cisco introduced the **object-group** on Cisco ASA Firewalls (and also on IOS routers since IOS 12.4.20T).

Purpose:

Network objects are designed to make firewall configuration easier and more manageable.

What they represent:

They can represent a single IP address, a range of IP addresses, a subnet (CIDR notation), or a fully qualified domain name (FQDN).

How they are used:

Once defined, network objects can be used in various configurations, such as access control lists (ACLs), Network Address Translation (NAT) rules, and service policies.

Benefits:

- **Simplified Configuration:** Instead of typing IP addresses or subnets multiple times, you can simply use the object name.
- **Centralized Management:** If you need to change an IP address or subnet, you only need to modify the object definition, and the change will be reflected everywhere it's used.
- **Improved Readability:** Using descriptive object names makes the configuration easier to understand and maintain.

Types of Network Objects:

- **Host:** A single IP address.
- **Network:** A subnet or range of IP addresses.
- **FQDN:** A fully qualified domain name.

Example:

Imagine you have a group of servers with IP addresses in the 192.168.1.0/24 network. Instead of specifying this subnet in every access rule, you can create a network object named "WebServers" and use that object in the rules.

Object Groups:

You can also group multiple network objects into object groups, which can further simplify configuration and management.

Examples of object-group:

- **Icmp-type** can be used to select all the different ICMP types, for example, echo, echo-reply, traceroute, unreachable, etc.
- **The network** is used to select IP addresses and/or network addresses.
- **Protocol** lets you select an entire protocol. For example, TCP, UDP, GRE, ESP, AH, OSPF, EIGRP, and many others.
- **Security** is used for Cisco TrustSec.
- **Service** is used to select TCP and/or UDP port numbers.
- **The user** is to select local user groups for Identity Firewall.

Service Objects in Cisco ASA Firewall

On a Cisco ASA firewall, a service object defines a specific protocol and port combination, which can then be used in access control lists (ACLs) and other security configurations, simplifying rule creation and maintenance.

Example:

A service object named "Web" could represent the TCP protocol on port 80 (HTTP).

Purpose:

Service objects are reusable components that represent a specific service or protocol (like HTTP, SSH, or FTP) and its associated port(s).

SERVICE GROUPS:

Concept:

You can also create service groups, which are collections of service objects.

Example:

A "Web Services" group could contain service objects for HTTP (port 80), HTTPS (port 443), and potentially other web-related services.

Benefits:

Service groups further simplify ACLs by allowing you to refer to a group of services instead of individual ones.

Project Task 2: Use the same topology to create a network object group for the Webservers

Configure>firewall>object>network object

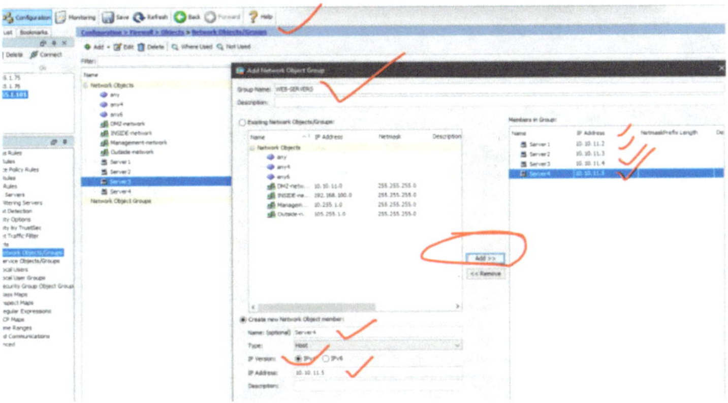

Project Task 3: Use the same topology to create a Service object group for the Webservers

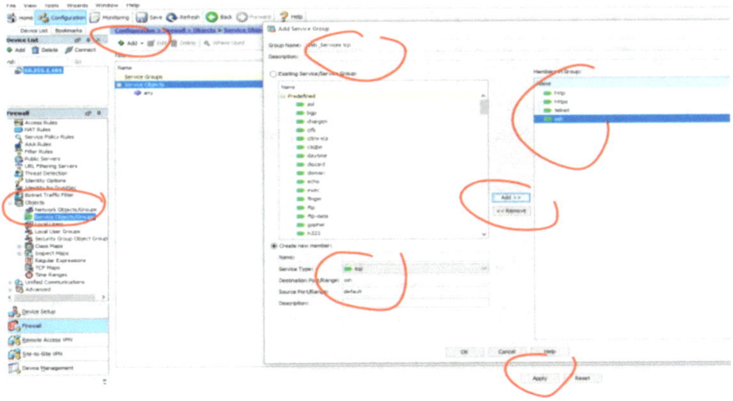

Project Task4: Configure a firewall policy allowing all users from the internet to access all the DMZ web servers on the following ports 80,443

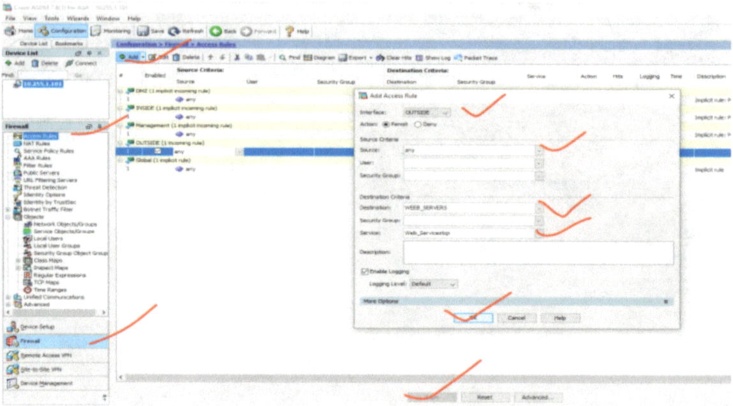

Now let's test if the ACL is working.

```
Internet#
Internet#telnet 10.10.11.2 80
Trying 10.10.11.2, 80 ... Open
X
HTTP/1.1 400 Bad Request
Date: Fri, 14 Mar 2025 17:35:18 GMT
Server: cisco-IOS
Accept-Ranges: none

400 Bad Request
[Connection to 10.10.11.2 closed by foreign host]
Internet#telnet 10.10.11.2 443
Trying 10.10.11.2, 443 ... Open
X
```

ASA ACL Project Using CLI

ACL -Lab Project _1

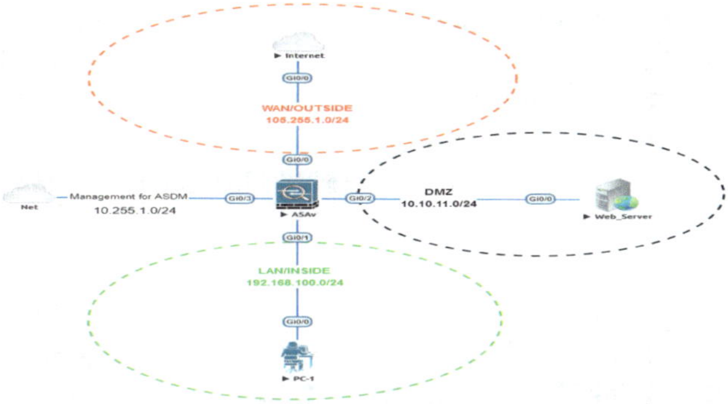

Project Task 1: **Configure a firewall policy that will permit any source outside (any IP address) to access the web server 10.10.11.2 on TCP port 80 (WWW) using CLI**

1. Apply basic Config on Cisco ASA Boston Office using CLI

- **Host name** – configure the hostname that provides identity for the ASA device

    Ciscoasa> en
    Password: hit enter to go to privileged mode
    Ciscoasa# conf t
    Ciscoasa(config)# **hostname ASA-FW1**

- **Enable password** - configures an enable password for the Cisco ASA that is required for users to access the privileged mode to configure all ASA features.

ASA-FW1 (config)# **enable password nptc123**

- **Timezone:** configures the time zone and daylight-saving time the firewall will use for logging and other events.

    ASA-FW1 (config) # **clock timezone EST -5 0**

ASA-FW1 (config) # **clock summer-time EST recurring**

- **Username and Password:** adds a user account with a username, password, and privilege level. Privilege level 15 allows our user to log in directly into the enabled mode of the ASA to perform configuration changes.
  ASA-FW1 (config) #**username admin password cisco privilege 15**

2. Configure AAA to use the local ASA database for telnet and SSH user authentication

ASA-FW1 (config) **# aaa authentication ssh console LOCAL**

ASA-FW1 (config) **# aaa authentication telnet console LOCAL**

3. Configure your inside, outside, and DMZ networks

ASA-FW1 (config) # **Int g0/0**
ASA-FW1 (config) # **Des link to Internet**
ASA-FW1 (config) # Nameif outside
INFO: Security level for ''outside" set to 0 by default
ASA-FW1 (config) # **Security-level 0**
ASA-FW1 (config) # **ip add 105.255.1.1 255.255.255.0**
ASA-FW1 (config) # **No shut**

ASA-FW1 (config) # **Int g0/1**
ASA-FW1 (config)# **Des LAN**
ASA-FW1 (config) # **Nameif Inside**
INFO: Security level for ''inside" set to 100 by default
ASA-FW1 (config) # **Security-level 100**
ASA-FW1 (config) # **Ip add 192.168.100.1 255.255.255.0**
ASA-FW1 (config) # **No shut**

Int

ASA-FW1 (config) # **Int g0/2**
ASA-FW1 (config) # **Des DMZ-Servers**
ASA-FW1 (config) # **Nameif DMZ**
ASA-FW1 (config) # **security-level 50**
ASA-FW1 (config) # **ip add 10.10.11.1 255.255.255.0**
ASA-FW1 (config) # **no shut**

```
Interface              IP-Address      OK? Method Status          Protocol
GigabitEthernet0/0     105.255.1.1     YES manual up              up
GigabitEthernet0/1     192.168.100.1   YES manual up              up
GigabitEthernet0/2     10.10.11.1      YES manual up              up
```

```
ASA-FW1# show nameif
Interface              Name                    Security
GigabitEthernet0/0     OUTSIDE                   0
GigabitEthernet0/1     INSIDE                  100
GigabitEthernet0/2     DMZ                      50
```

4. Configure telnet

To enable Telnet, we will specify which hosts or subnets can Telnet into the ASA, including the interface to which we can Telnet. As a best practice, Telnet will only be enabled on our inside interface for any computer on the inside subnet (192.168.100.0) to access the ASA via Telnet. Telnet 192.168.100.0 255.255.252.0 inside

ASA-FW1 (config) # **telnet 192.168.100.0 255.255.255.0 inside**

5. Configure SSH

Next, we will enable SSH the same way we enabled telnet, but first, we need to configure the domain name and generate our local RSA keys (1024 bits).

ASA-FW1 (config) # **domain-name nptc.com**
ASA-FW1 (config) # **crypto key generate rsa modulus 1024**

NB: Once our RSA keys have been generated, we will enable SSH access for any host from the inside network (192.168.100.0) and a host located outside the network (105.255.1.0).

ASA-FW1 (config)**# ssh 105.255.1.101 255.255.255.0 outside**
ASA-FW1 (config) # **ssh 192.168.100.0 255.255.255.0 inside**

6. Configure a firewall policy that will permit any source (any IP address) to access the web server 10.10.11.2 on TCP port 80 (WWW)

Now let's test if port 80 is open on the internet.

```
Internet#telnet 10.10.11.2 80
Trying 10.10.11.2, 80 ...
% Connection timed out; remote host not responding
```

This traffic is denied by default because it is coming from a low Security Zone to a high Security Zone. Let's create an access list that allows HTTP traffic. We'll create something so that users on the internet are allowed to connect to the web server on port 80. All other traffic will be denied.

ASA-FW1 (config)# **access-list OUTSIDE_DMZ extended permit tcp any host 10.10.11.2 eq 80**

ASA-FW1 (config)# **access-group OUTSIDE_DMZ in interface OUTSIDE**

Let's verify if the ACL is working

```
Internet#telnet 10.10.11.2 80
Trying 10.10.11.2, 80 ... Open
```

```
ASA-FW1(config)# show access-list
access-list cached ACL log flows: total 0, denied 0 (deny-flow-max 4096)
            alert-interval 300
access-list OUTSIDE_DMZ; 1 elements; name hash: 0xf68adb0a
access-list OUTSIDE_DMZ line 1 extended permit tcp any host 10.10.11.2 eq www (hitcnt=1) 0xdba32ba0
```

Global Access-List

The global access-list is useful when you have many interfaces and you don't want to enable an access-list on each one of them. When you use this, you create an access list like you normally do, but instead of enabling it on an interface, we enable it globally.

When you do this, the access list is applied to **all outbound traffic on all interfaces**. It doesn't work for outbound traffic.

Texas-ASAFW (config)# no **access-group OUTSIDE_DMZ in interface OUTSIDE**

Texas-ASAFW (config)# **access-group OUTSIDE_DMZ global**

And it will achieve the same result

```
Internet#telnet 10.10.11.2 80
Trying 10.10.11.2, 80 ... Open
```

Now I will create a network object group using the host IP addresses of mail servers at the DMZ:

ASA-FW1 (config) # **object-group network WEB_SERVERS**

ASA-FW1 (config-network-object-group) # **network-object host 10.10.11.2**

ASA-FW1 (config-network-object-group) # **network-object host 10.10.11.3**

ASA-FW1 (config-network-object-group) # **network-object host 10.10.11.4**

ASA-FW1 (config-network-object-group) # **network-object host 10.10.11.5**

ASA-FW1 (config-network-object-group) # **network-object host 10.10.11.5**

```
ASA-FW1(config)# show run object-group
object-group network WEB_SERVERS
 network-object host 10.10.11.2
 network-object host 10.10.11.3
 network-object host 10.10.11.4
 network-object host 10.10.11.5
 network-object host 10.10.11.6
```

Now let's configure the firewall policy or ACL using the object group to allow users from outside access the internet

ASA-FW1 (config)# **access-list HTTP_TO_DMZ permit tcp any object-group WEB_SERVERS eq 80**

Without the object group, the ACL will have 5 lines instead of the above

```
ASA-FW1(config)# show access-list
access-list cached ACL log flows: total 0, denied 0 (deny-flow-max 4096)
            alert-interval 300
access-list HTTP_TO_DMZ; 5 elements; name hash: 0x6ce713ae
access-list HTTP_TO_DMZ line 1 extended permit tcp any object-group WEB_SERVERS eq www (hitcnt=0) 0x0964f55b
  access-list HTTP_TO_DMZ line 1 extended permit tcp any host 10.10.11.2 eq www (hitcnt=0) 0xbe25da03
  access-list HTTP_TO_DMZ line 1 extended permit tcp any host 10.10.11.3 eq www (hitcnt=0) 0xbace7e6d
  access-list HTTP_TO_DMZ line 1 extended permit tcp any host 10.10.11.4 eq www (hitcnt=0) 0x303325f3
  access-list HTTP_TO_DMZ line 1 extended permit tcp any host 10.10.11.5 eq www (hitcnt=0) 0x29dcf20a
  access-list HTTP_TO_DMZ line 1 extended permit tcp any host 10.10.11.6 eq www (hitcnt=0) 0x967a574d
```

We can also create a Service object group that combines all our TCP ports. E.g., ports 22, 23, and 80,443, and apply it to the web servers we have at the DMZ.

ASA-FW1 (config) # **object-group service WEB_SERVICES tcp**

ASA-FW1 (config-service-object-group) # **port-object eq 22**

ASA-FW1 (config-service-object-group) # **port-object eq 23**

ASA-FW1 (config-service-object-group) # **port-object eq 80**

ASA-FW1 (config-service-object-group) # **port-object eq 443**

```
ASA-FW1(config-service-object-group)# show run object-group
object-group network WEB_SERVERS
 network-object host 10.10.11.2
 network-object host 10.10.11.3
 network-object host 10.10.11.4
 network-object host 10.10.11.5
 network-object host 10.10.11.6
object-group service DMZ_WEB_SERVICES tcp
 port-object eq ssh
 port-object eq telnet
 port-object eq www
 port-object eq https
```

Let's now configure the ACL using both the network object group and the service object group

ASA-FW1 (config) # **access-list HTTP_TO_DMZ permit tcp any object-group WEB_SERVERS object-group DMZ_WEB_SERVICES**

Let's apply the ACL to an interface

ASA-FW1 (config) # **access-group HTTP_TO_DMZ global**

```
ASA-FW1(config)# show access-list
access-list cached ACL log flows: total 0, denied 0 (deny-flow-max 4096)
            alert-interval 300
access-list HTTP_TO_DMZ; 20 elements; name hash: 0x6ce713ae
access-list HTTP_TO_DMZ line 1 extended permit tcp any object-group WEB_SERVERS object-group DMZ_WEB_SERVICES (hitcnt=0) 0x25002ebc
  access-list HTTP_TO_DMZ line 1 extended permit tcp any host 10.10.11.2 eq ssh (hitcnt=0) 0x588491df
  access-list HTTP_TO_DMZ line 1 extended permit tcp any host 10.10.11.2 eq telnet (hitcnt=0) 0x8fa088ea
  access-list HTTP_TO_DMZ line 1 extended permit tcp any host 10.10.11.2 eq www (hitcnt=0) 0xbe25da03
  access-list HTTP_TO_DMZ line 1 extended permit tcp any host 10.10.11.2 eq https (hitcnt=0) 0x624a41a8
  access-list HTTP_TO_DMZ line 1 extended permit tcp any host 10.10.11.3 eq ssh (hitcnt=0) 0x6a506109
  access-list HTTP_TO_DMZ line 1 extended permit tcp any host 10.10.11.3 eq telnet (hitcnt=0) 0x26f11125
  access-list HTTP_TO_DMZ line 1 extended permit tcp any host 10.10.11.3 eq www (hitcnt=0) 0xbace7e6d
  access-list HTTP_TO_DMZ line 1 extended permit tcp any host 10.10.11.3 eq https (hitcnt=0) 0xf49f500a
  access-list HTTP_TO_DMZ line 1 extended permit tcp any host 10.10.11.4 eq ssh (hitcnt=0) 0x6367c289
  access-list HTTP_TO_DMZ line 1 extended permit tcp any host 10.10.11.4 eq telnet (hitcnt=0) 0x587077f8
  access-list HTTP_TO_DMZ line 1 extended permit tcp any host 10.10.11.4 eq www (hitcnt=0) 0x303325f3
  access-list HTTP_TO_DMZ line 1 extended permit tcp any host 10.10.11.4 eq https (hitcnt=0) 0x6653c3c2
  access-list HTTP_TO_DMZ line 1 extended permit tcp any host 10.10.11.5 eq ssh (hitcnt=0) 0x569d9d04
  access-list HTTP_TO_DMZ line 1 extended permit tcp any host 10.10.11.5 eq telnet (hitcnt=0) 0xab6d8651
  access-list HTTP_TO_DMZ line 1 extended permit tcp any host 10.10.11.5 eq www (hitcnt=0) 0x29dcf20a
  access-list HTTP_TO_DMZ line 1 extended permit tcp any host 10.10.11.5 eq https (hitcnt=0) 0x07087f43
  access-list HTTP_TO_DMZ line 1 extended permit tcp any host 10.10.11.6 eq ssh (hitcnt=0) 0x7b8c9500
  access-list HTTP_TO_DMZ line 1 extended permit tcp any host 10.10.11.6 eq telnet (hitcnt=0) 0xfd4356a2
  access-list HTTP_TO_DMZ line 1 extended permit tcp any host 10.10.11.6 eq www (hitcnt=0) 0x967a574d
  access-list HTTP_TO_DMZ line 1 extended permit tcp any host 10.10.11.6 eq https (hitcnt=0) 0x2c3e6a5f
```

Cisco ASA Remove Access-List

If you want to remove an access list from a Cisco ASA Firewall, then you'll find out that removing it doesn't work the same as on Cisco IOS routers or switches. Let me give you an example of creating an access-list and then try to remove it:

```
ASA-FW1(config)# no access-list OUTSIDE_DMZ
ERROR: % Incomplete command
```

Using "no" in front of it doesn't work; the ASA thinks that we want to remove a single entry, not delete the entire access-list. The following command will work, or you can apply no using the complete ACL command

ASA-FW1 (config) # **clear configure access-list OUTSIDE_DMZ**

Use the **clear configure** command to get rid of the entire access list. Let's verify this:

```
ASA-FW1(config)# clear  configure access-list OUTSIDE_DMZ
ASA-FW1(config)# show run ac
ASA-FW1(config)# show run access-l
ASA-FW1(config)#
```

ASA-FW1 (config) # show run access-g

NB: This configuration will also affect the access-group interface and should not be done on production

Cisco ASA ACL Best Practices

Always apply ACLs inbound on all interfaces

I don't like to apply ACLs outbound on the interfaces because I want to use the firewall's internal compute and memory resources as efficiently as possible.

Name the ACL after the interface on which the rule will be applied

E.g. OUTSIDE-to-DMZ, inside-to-in, https-to-DMZ

Use remarks in your ACLs to internally document your intentions

The more you can make the configuration of your firewall self-documenting, the easier it will be to manage it going forward

Use object groups

For example, I might want to block a particular set of malicious IP addresses from ever accessing my network from the outside. If I use the same object group on the inside interface, I can also prevent anybody inside my network from ever accessing these same malicious external hosts. And if I add a new host to that object group, I automatically update both those inbound and outbound rules.

Only use object groups when you have several TCP/UDP ports or source/destination addresses that need to be grouped. Always using object-groups even for rules that have a low number of services or network defined might eventually make the configuration harder to read.

Make your ACL as specific as possible.

Don't permit "any" hosts if you can narrow it down. Make those "permit" rules as specific as possible. The same goes for protocols. Don't permit all IP protocols if you really mean a particular protocol. So don't undermine your security.

Generally, I like to build my ACLs in a structured way. First, I include a relatively small and very specific whitelist. It includes things that I know are always allowed, and overrides any blacklist rules that might come later.

The blacklist

My general blacklist is usually a list of sites or IP address ranges representing geographic regions that I will never accept anything.

Cisco ASA NAT Configuration

Network Address Translation (NAT) is a service that enables private IP networks to use the Internet and cloud. NAT translates private IP addresses in an internal network to a public IP address before packets are sent to an external network.

Network Address Translation (NAT) is a service that operates on an edge (Router, Firewall) to connect private networks to public networks like the internet. NAT is often implemented at the WAN edge to enable internet access in core, campus, branch, and colocation sites

How it Works

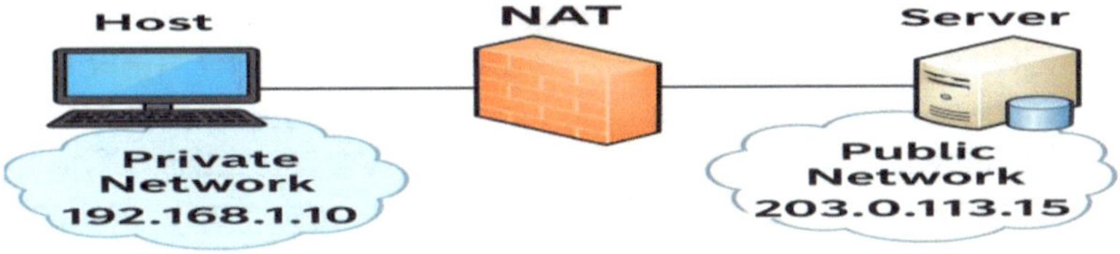

What are the Situations where Nat is required?

1. When we need to connect to the internet and our host doesn't have a globally unique IP address.
2. When we want to hide Internal IP addresses from outside for security purposes.
3. A company is going to merge with another company that uses the same address space.

Advantages of NAT

- NAT conserves legally registered IP addresses.
- It provides privacy as the device's IP address, sending and receiving the traffic, will be hidden.
- Eliminates address renumbering when a network evolves.
- Nat also allows individuals and organizations to establish cost cost-effective and simple connection.
- NAT prevents IP address overlapping.

Disadvantages of NAT

- Translation results in switching path delays.
- Certain applications will not function while NAT is enabled.
- Complicates tunneling protocols such as IPsec.
- Also, the router, being a network layer device, should not tamper with port numbers (transport layer), but it has to do so because of NAT.

Network Address Translation (NAT) Types

There are 3 ways to configure NAT:

Static NAT – In this, a single unregistered (Private) IP address is mapped with a legally registered (Public) IP address, i.e, one-to-one mapping between local and global addresses. This is generally used for Web hosting. These are not used in organizations as there are many devices that will need Internet access, and to provide Internet access, a public IP address is needed.

Dynamic NAT – In this type of NAT, an unregistered IP address is translated into a registered (Public) IP address from a pool of public IP addresses. If the IP address of the pool is not free, then the packet will be dropped, as only a fixed number of private IP addresses can be translated to public addresses.
Suppose that there is a pool of 2 public IP addresses; then, only 2 private IP addresses can be translated at a given time. If a 3rd private IP address wants to access the Internet, then the packet will be dropped; therefore, many private IP addresses are mapped to a pool of public IP addresses. NAT is used when the number of users who want to access the Internet is fixed. This is also very costly, as the organization has to buy many global IP addresses to make a pool.

Port Address Translation (PAT) – This is also known as NAT overload. In this, many local (private) IP addresses can be translated to a single registered IP address. Port numbers are used to distinguish the traffic, i.e., which traffic belongs to which IP address. This is most frequently used as it is cost-effective, as thousands of users can be connected to the Internet by using only one real global (public) IP address.

Dynamic Nat Project

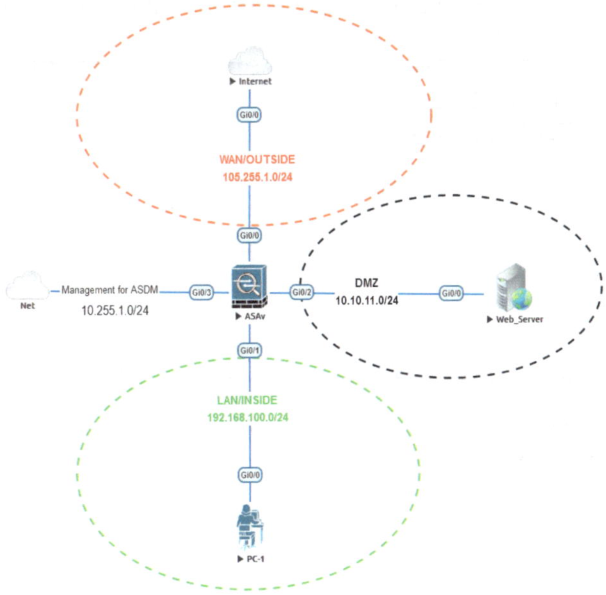

Network Object

In ASA, every configuration of NAT requires an object. When a packet enters the ASA, both the source and destination IP addresses are checked against the network object NAT rules. Object represents any one item.

- **Network Object**- Represents a single **IP address, Subnet, or Range**
- **Service Object** – Represents a single Port/Protocol

  Port- eq=equal,gt=greater than,lt=less than, neq=not equal, range
  Protocol-tcp, udp, icmp, gres,esp etc

Dynamic Nat Project

1. **Configure the network object IP pool to be use for dynamic NAT for both the public pool and DMZ pool**

ASA-FW1(config)# object network Public_Pool

ASA-FW1(config-network-object)# range 105.255.1.100 105.255.1.200

ASA-FW1(config-network-object)# object network DMZ_Pool

ASA-FW1(config-network-object)# range 10.10.11.100 10.10.11.200

2. **Configure dynamic NAT for users over the internet to access the web-facing Application of Web facing on port 23**

ASA-FW1(config)# object network DMZ_TO_OUTSIDE
ASA-FW1(config-network-object)# subnet 10.10.11.0 255.255.255.0
ASA-FW1(config-network-object)# nat (DMZ, outside) dynamic Public_Pool

3. Generate traffic from the DMZ to the internet to see if their IP packets are correctly translated.

Web_Server#**telnet 105.255.1.2 23**
Trying 105.255.1.2 ... Open

```
*****************************************************************************
* IOSv is strictly limited to use for evaluation, demonstration, and IOS *
* education. IOSv is provided as-is and is not supported by Cisco's      *
* Technical Advisory Center. Any use or disclosure, in whole or in part, *
* of the IOSv Software or Documentation to any third party for any       *
* purposes are expressly prohibited except as otherwise authorized by    *
* Cisco in writing.                                                      *
*****************************************************************************
```

User Access Verification

Password:

We have a connection, so let's see if we have a translation:

ASA-FW1(config-network-object) # **show nat**

Auto NAT Policies (Section 2)
1 (DMZ) to (outside) source dynamic DMZ_TO_OUTSIDE Public_Pool
translate_hits = 1, untranslate_hits = 0
ASA-FW1(config-network-object) # show nat

Auto NAT Policies (Section 2)
1 (DMZ) to (outside) source dynamic DMZ_TO_OUTSIDE Public_Pool
translate_hits = 2, untranslate_hits = 0

ASA-FW1(config-network-object) # **show nat de**

Auto NAT Policies (Section 2)
1 (DMZ) to (outside) source dynamic DMZ_TO_OUTSIDE Public_Pool
translate_hits = 2, untranslate_hits = 0
Source - Origin: 10.10.11.0/24, Translated: 105.255.1.100-105.255.1.200

ASA-FW1(config-network-object) #**show xlate**
1 in use, 1 most used
Flags: D - DNS, e - extended, I - identity, i - dynamic, r - portmap,
s - static, T - twice, N - net-to-net

NAT from DMZ: 10.10.11.2 to outside: 105.255.1.157 flags i idle 0:05:01 timeout 3:

4. **Remove the following and use ASDM to configure the same**

   a. NAT config –under nat rules
   b. Nat object – under object

5. **Apply ASDM configuration to configure the same by removing the nat configuration on cli**

Using ASDM to create a dynamic NAT

Step 1: Create your public object pool

Configuration > firewall > object> Add > add network object

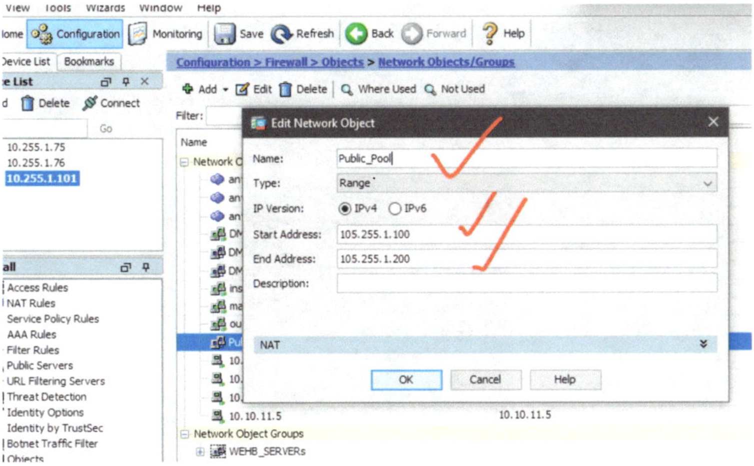

Step 2 Configure the Nat object for the outside

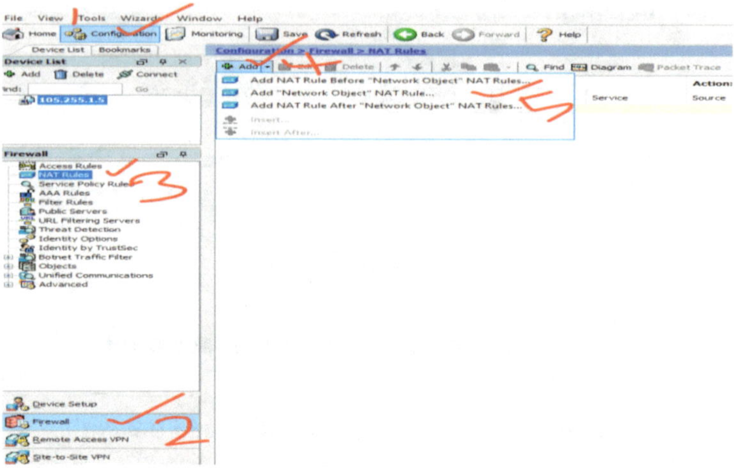

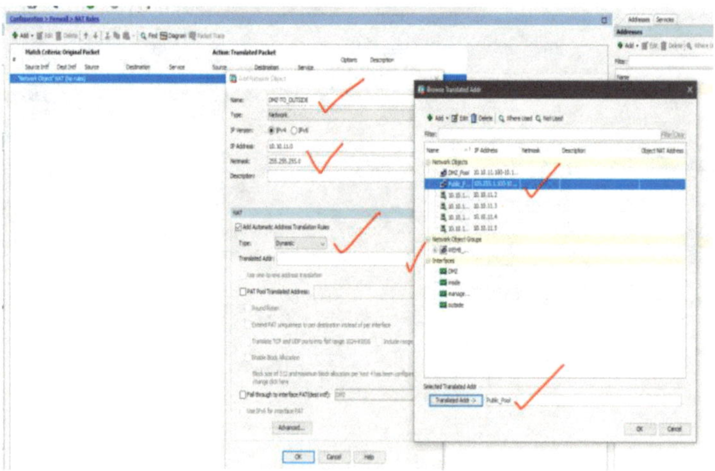

Step 3: Apply the interface for the NAT

Click on the Nat rule > click on advance > apply on the right interface

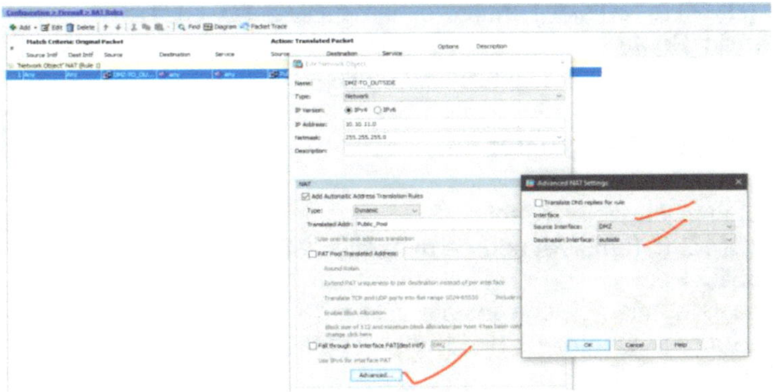

Step 4. Generate traffic from the internet to the DMZ using Packet Tracer to test if the NAT is working.

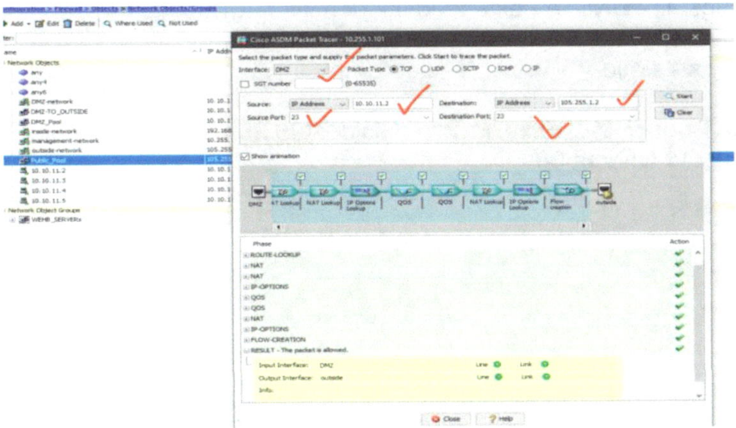

Verify on the CLI

ASA-FW1# **show nat**

Auto NAT Policies (Section 2)
1 (DMZ) to (outside) source dynamic DMZ-TO_OUTSIDE Public_Pool
translate_hits = 1, untranslate_hits = 0

ASA-FW1# **show nat de**

Auto NAT Policies (Section 2)
1 (DMZ) to (outside) source dynamic DMZ-TO_OUTSIDE Public_Pool
translate_hits = 1, untranslate_hits = 0
Source - Origin: 10.10.11.0/24, Translated: 105.255.1.100-105.255.1.200

ASA-FW1# **show xlate**
1 in use, 1 most used
Flags: D - DNS, e - extended, I - identity, i - dynamic, r - portmap,
s - static, T - twice, N - net-to-net

NAT from DMZ:10.10.11.2 to outside:105.255.1.105 flags i idle 0:01:11 timeout 30

PAT (NAT Overload)

PAT is primarily required when LAN users are translated to a public IP (interface IP or IP from Public Pool). This type of **Dynamic NAT/PAT** configuration is used to provide internet access to LAN Users by translating the LAN Subnet with the Outside Interface of the Firewall or any Public IP address.

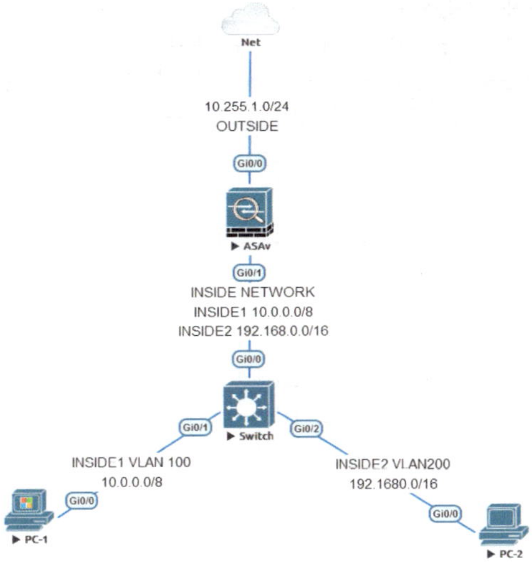

1. Configuring ASDM access using the OUTSIDE interface for management and using the ASDM to create your LAN

ciscoasa(config)# int g0/0
ciscoasa(config-if)# des link to the Internet
ciscoasa(config-if)# nameif outside
ciscoasa(config-if)# security-level 0
ciscoasa(config-if)# ip add 10.255.1.201 255.255.255.0
ciscoasa(config-if)# no shut
ciscoasa(config-if)#
ciscoasa(config-if)#
ciscoasa(config-if)# http server enable
ciscoasa(config)# http 10.255.1.0 255.255.255.0 outside
ciscoasa(config)# username admin pass cisco

NB: With this configuration, we can now access the ASDM

2. Using the ASDM to Create Interfaces for the LAN

Configure >Device setup>interface settings > interfaces >add

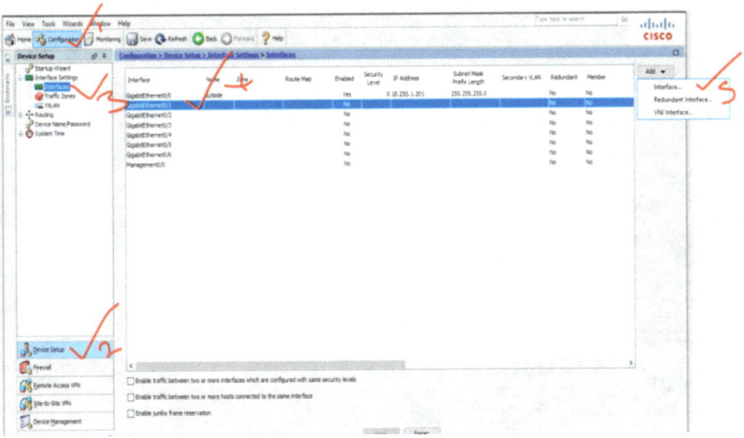

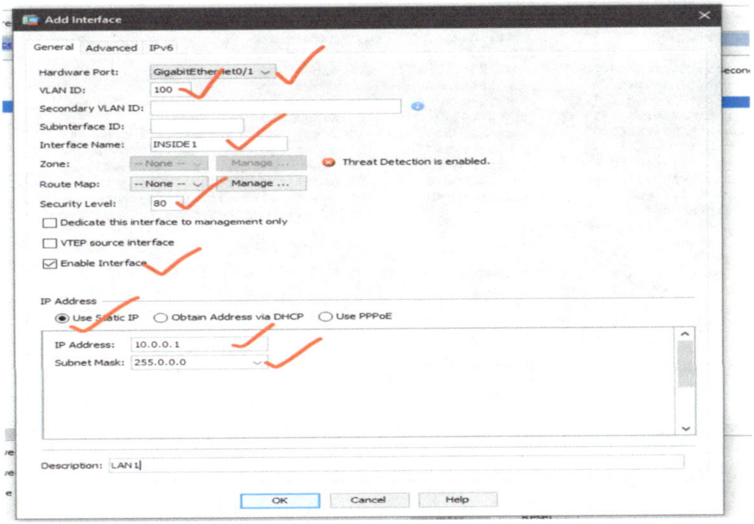

3. **Repeat the same for the second sub-interface using VLAN 200**

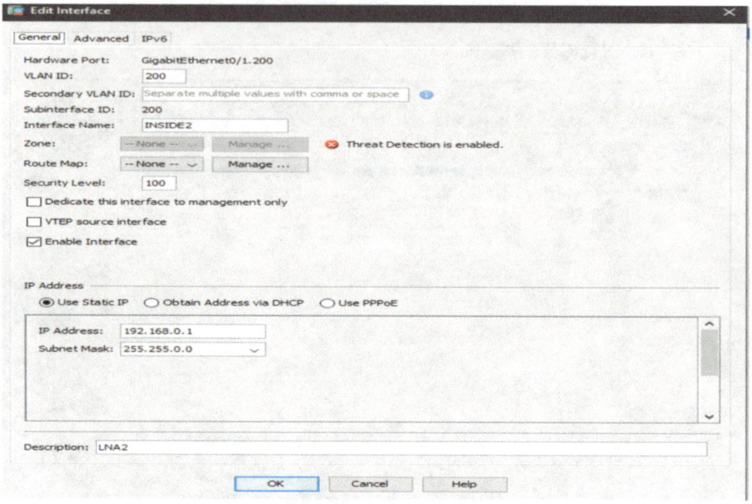

Verify with the command line.

```
ASA-FW1(config-if)# show int ip br
Interface                  IP-Address      OK? Method Status                Prot
ocol
GigabitEthernet0/0         10.255.1.101    YES manual up                    up
GigabitEthernet0/1         unassigned      YES unset  up                    up
GigabitEthernet0/1.100     10.0.0.1        YES manual up                    up
GigabitEthernet0/1.200     192.168.0.1     YES manual up                    up
GigabitEthernet0/2         unassigned      YES unset  administratively down up
GigabitEthernet0/3         unassigned      YES unset  administratively down up
GigabitEthernet0/4         unassigned      YES unset  administratively down up
GigabitEthernet0/5         unassigned      YES unset  administratively down up
GigabitEthernet0/6         unassigned      YES unset  administratively down up
Management0/0              unassigned      YES unset  administratively down up
ASA-FW1(config-if)# show nameif
Interface                Name                     Security
GigabitEthernet0/0       outside                    0
GigabitEthernet0/1.100   INSIDE1                   80
GigabitEthernet0/1.200   INSIDE2                  100
```

4. Configure the default route using ASDM

Configure >device setup>static route >add

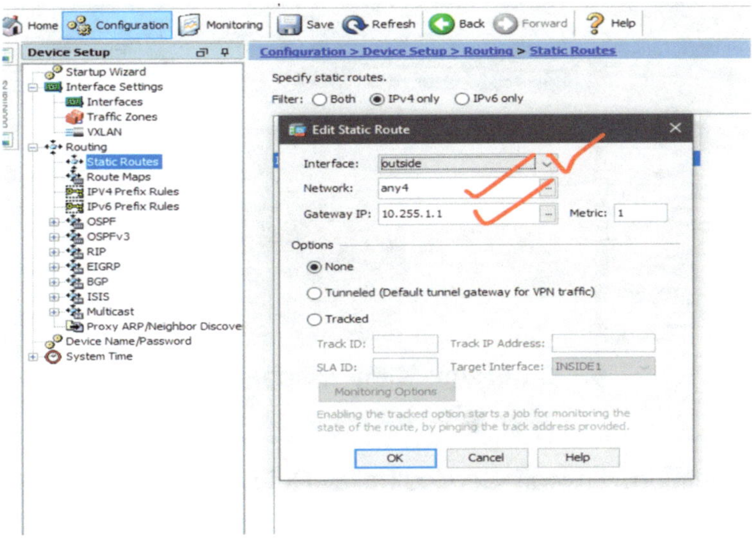

5. Configure PAT as your NAT protocol to give internet access to your inside network.

Let's create for inside1 on network of 10.0.0.0/16

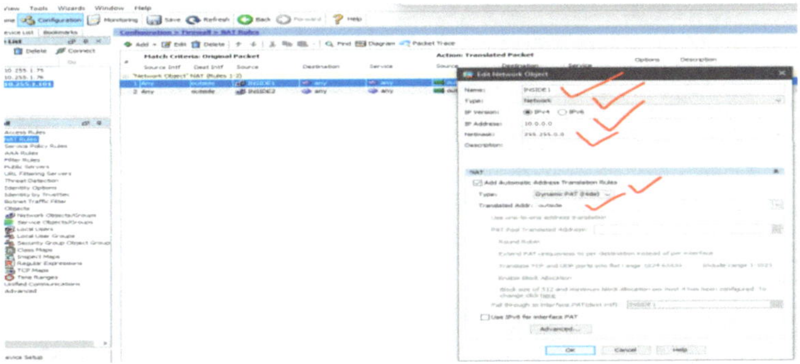

Let's create for inside2 on network of 192.168.0.0/16

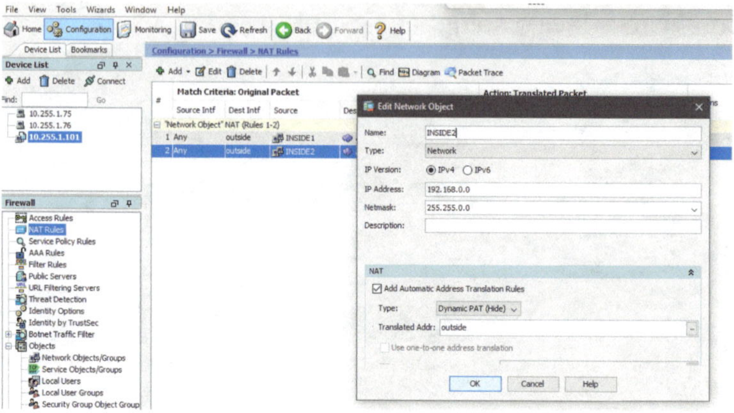

6. **Configure ICMP inspection on ASDM to allow ping access.**

Configure>firewall>Service Policy Rules

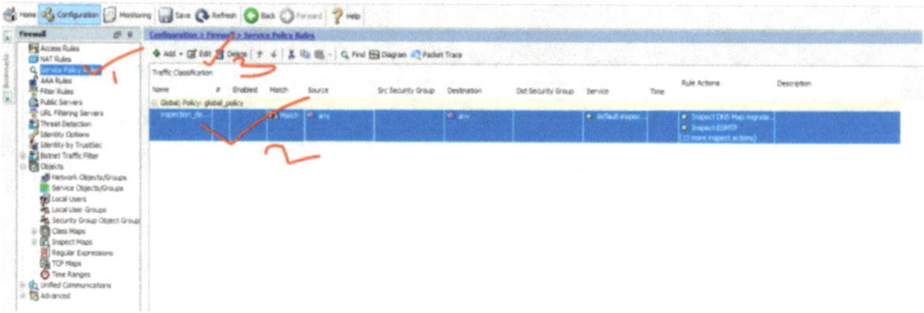

1. Select Rule Actions
2. Click Protocol Inspection
3. Check "ICMP" and "ICMP Error" and hit OK

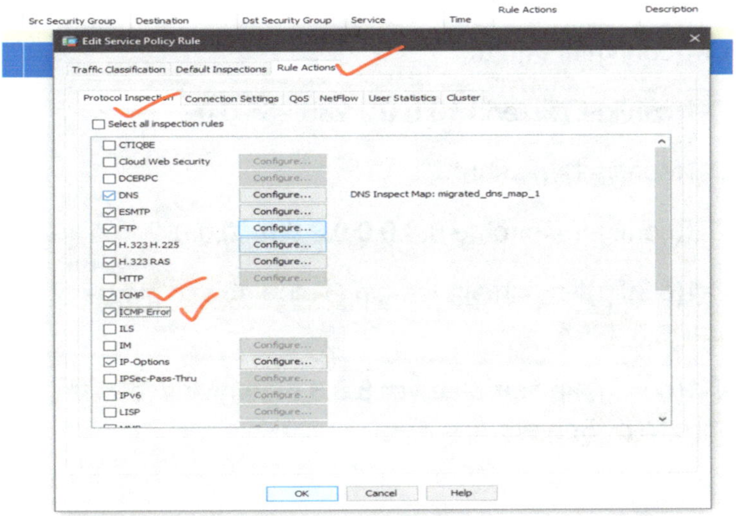

Now we can verify on both PCs for internet access.

Configurations to be done on the switch

SW1(config)#**interface g0/0**

SW1(config-if)#**switchport trunk encapsulation dot1q**

SW1(config-if)#**switchport mode trunk**

SW1(config)#**interface g0/1**

SW1(config-if)#**switchport mode access**

SW1(config-if)#**switchport access vlan 100**

Router Users Configuration As Seen On the Topology

PC-1(config)#hostname PC1

PC-1(config)#int g0/0

PC-1(config-if)#ip add 10.0.0.2 255.255.0.0

PC-1(config-if)#no shut

PC-1(config-if)#ip route 0.0.0.0 0.0.0.0 10.0.0.1

PC-1(config)#ip domain-lookup (To be able to ping a domain name)

PC-1(config)#ip name-server 8.8.8.8 (to give access to public DNS Server)

PC1(config)#**do ping 8.8.8.8**

Type escape sequence to abort.

Sending 5, 100-byte ICMP Echos to 8.8.8.8, timeout is 2 seconds:

!!!!!

Success rate is 100 percent (5/5), round-trip min/avg/max = 30/32/34 ms

PC1(config)#**do ping facebook.com**

Type escape sequence to abort.

Sending 5, 100-byte ICMP Echos to 31.13.66.35, timeout is 2 seconds:

!!!!!

Success rate is 100 percent (5/5), round-trip min/avg/max = 27/34/41 ms

PC2 Configuration

PC-2(config)#**int g0/0**

PC-2(config-if)#**ip add 192.168.0.2 255.255.0.0**

PC-2(config-if)#**no shut**

PC-2(config-if)#**ip route 0.0.0.0 0.0.0.0 192.168.0.1**

PC-2(config)#ip domain-lookup (To be able to ping a domain name)

PC-2(config)#ip name-server 8.8.8.8 (to give access to public DNS Server)

PC2 (config)#do ping 8.8.8.8

Type escape sequence to abort.

Sending 5, 100-byte ICMP Echos to 8.8.8.8, timeout is 2 seconds:

!!!!!

Success rate is 100 percent (5/5), round-trip min/avg/max = 28/31/33 ms

PC2(config)#do ping ghanaweb.com

Type escape sequence to abort.

Sending 5, 100-byte ICMP Echos to 104.21.41.135, timeout is 2 seconds:

!!!!!

Success rate is 100 percent (5/5), round-trip min/avg/max = 28/33/41 ms

Let's now verify the NAT on the ASA

ASA-FW1 (config-if)# **show nat**

Auto NAT Policies (Section 2)

1 (any) to (outside) source dynamic INSIDE1 interface translate_hits = 5, untranslate_hits = 0

2 (any) to (outside) source dynamic INSIDE2 interface translate_hits = 3, untranslate_hits = 0

ASA-FW1 (config-if) #**show xlate**
1 in use, 3 most used
Flags: D - DNS, e - extended, I - identity, i - dynamic, r - portmap,
s - Static, T - twice, N - net-to-net

ICMP PAT from any:192.168.0.2/6 to outside:10.255.1.101/6 flags ri idle 0:00:04 timeout 0:00:30

STATIC NAT

Static NAT is primarily required when a Data Center or Hub site has a Web-facing server in the DMZ Zone or Inside Zone, and Users over the Internet need to access the Application of the Web-facing server. The applications may be **a Web (HTTP/HTTPS) Server, an Email Server, or even an FTP Server**. Below is a sample scenario where an Application server is hosted in the DMZ Zone and needs to be accessed from outside (Internet)

Project Task. Configure an inbound rule to allow users on the internet to connect to our DMZ E-Commerce Webserver (10.10.11.2) listening on port 443

Static NAT Project Lab

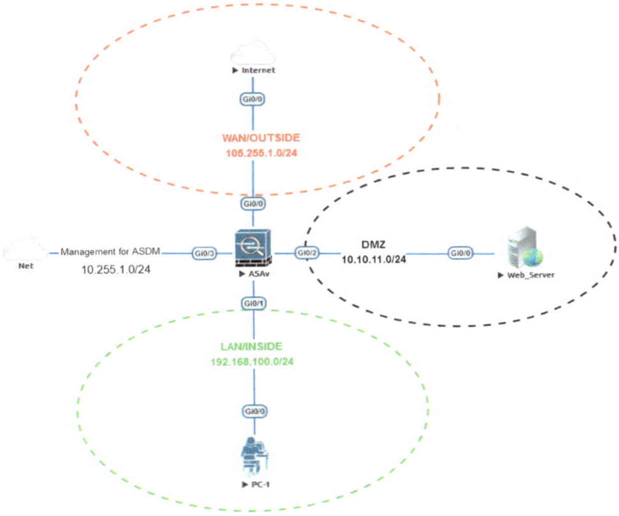

Step 1. First, we will create a network object that defines our "webserver" in the DMZ and also configure to what IP address it should be translated

ASA-1(config) # object network WEB-SERVER

ASA-1(config-network-object) # host 10.10.11.2

ASA-1(config-network-object) # nat (DMZ, OUTSIDE) static 105.255.1.200

NB: The configuration above tells the ASA that whenever an outside device connects to IP address 105.255.1.200, it should be translated to IP address 10.10.11.2. This takes care of NAT, but we still have to create an access list, or traffic will be dropped

ASA1(config)# **access-list OUTSIDE_TO_DMZ extended permit tcp any host 10.10.11.2 eq 443**

ASA1 (config)# **access-group OUTSIDE_TO_DMZ in interface OUTSIDE**

This enables the access-list on the outside interface

Internet#**telnet 105.255.1.200 443**

Trying 105.255.1.200, 443 ... Open

ASA-FW1(config)# **show access-list**

access-list cached ACL log flows: total 0, denied 0 (deny-flow-max 4096)

alert-interval 300

access-list outside_access_in; 1 elements; name hash: 0x6892a938

access-list outside_access_in line 1 extended permit tcp any object Web_Server eq https (hitcnt=1) 0xf2417cc4

ASA-FW1(config)# **show xlate**

1 in use, 1 most used

Flags: D - DNS, e - extended, I - identity, i - dynamic, r - portmap,

s - Static, T - twice, N - net-to-net

NAT from DMZ: 10.10.11.2 to outside: 105.255.1.200

Flags s idle 0:00:09 timeout 0:00:00

Apply the same configuration using ASDM to the same Project Task

Step 1: Create a network object that defines our "webserver" in the DMZ and also configure it to what IP address it should be translated.

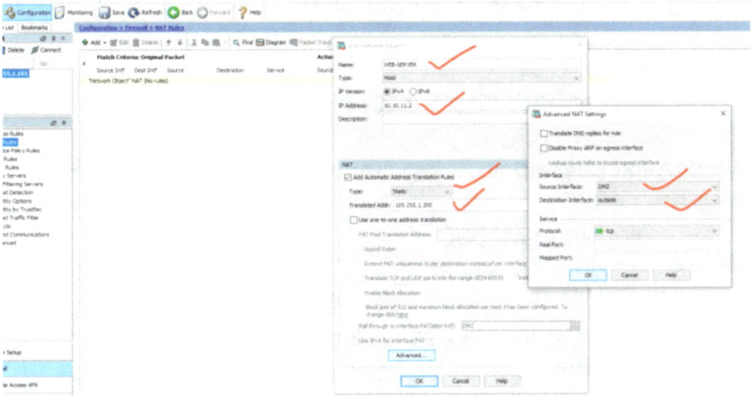

Step 2: Apply ACL to allow inbound traffic.

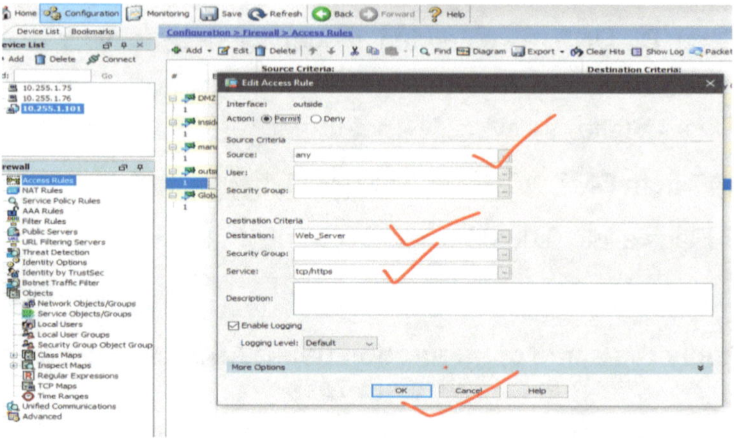

Internet#**telnet 105.255.1.200 443**
Trying 105.255.1.200, 443 ... Open

ASA-FW1(config)# **show access-list**

access-list cached ACL log flows: total 0, denied 0 (deny-flow-max 4096)
alert-interval 300
access-list outside_access_in; 1 elements; name hash: 0x6892a938
access-list outside_access_in line 1 extended permit tcp any object Web_Server eq https (hitcnt=1) 0xf2417cc4

ASA-FW1(config)# **show xlate**
1 in use, 1 most used
Flags: D - DNS, e - extended, I - identity, i - dynamic, r - portmap,
s - Static, T - twice, N - net-to-net
NAT from DMZ: 10.10.11.2 to outside: 105.255.1.200
Flags s idle 0:00:09 timeout 0:00:00

Static NAT Port Forwarding

NAT Port Forwarding is useful when you have a single public IP address and multiple devices behind it that you want to reach from the outside world

Project Task-1: Whenever someone connects to IP address 105.255.1.1 TCP port 80, we will forward them to 10.10.11.2 TCP port 80.

Step 1: We create a network object that specifies the real IP address of the web server, and then we create our NAT rule. By using the keyword **interface,** we tell the ASA to use the IP address on the (outside) interface. The first port number is the port that the server is **listening on**, and the second port number is the outside port number.

Using the command Line to configure

ASA-1(config) # object network WEB-SERVER

ASA-1(config-network-object) #host 10.10.11.2

ASA-1(config-network-object) # nat (DMZ, OUTSIDE) static interface service tcp 80 80

We can also use the keyword interface; we tell the ASA to use the IP address on the (outside) interface.

Step 2: Let's configure the Access List to allow traffic from the outside to the server

ASA-1(config)# **access-list out_DMZ extended permit tcp any host 192.168.3.1 eq 80**

ASA1(config)# **access-group out_DMZ in interface OUTSIDE**

Let's generate some traffic for testing

Internet#telnet 105.255.1.200 80

Trying 105.255.1.200, 80 ... Open

X

ASA-FW1(config)# show xlate

1 in use, 2 most used

Flags: D - DNS, e - extended, I - identity, i - dynamic, r - portmap,

s - static, T - twice, N - net-to-net

TCP PAT from DMZ: 10.10.11.2 80-80 to outside: 105.255.1.1 80-80

flags sr idle 0:00:18 timeout 0:00:00

Using ASDM to configure the same

Step 1: Creating the static NAT port forwarding on port 80 80

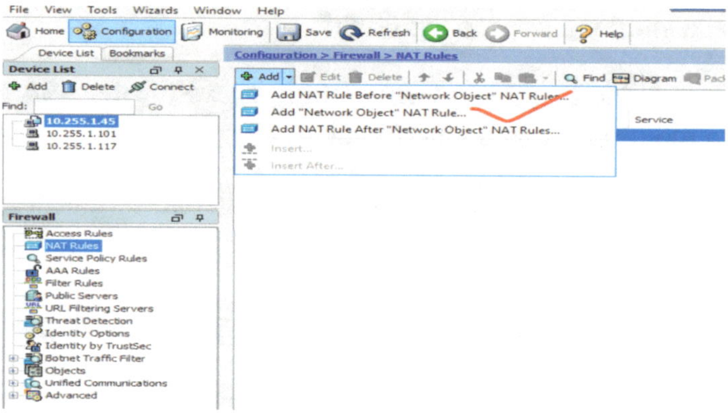

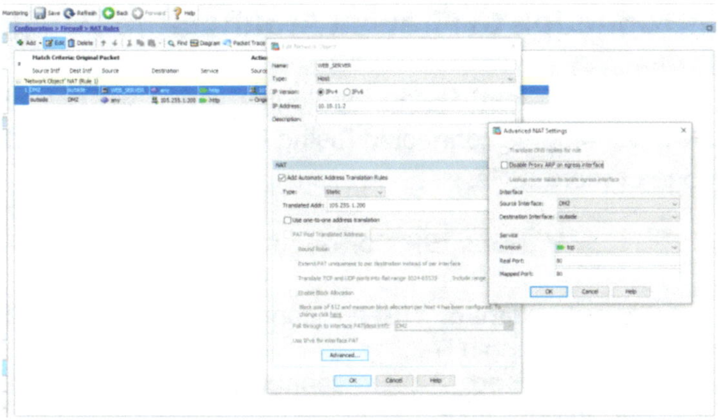

Let's generate traffic on the internet.

Internet#telnet 105.255.1.200 80
Trying 105.255.1.200, 80 ... Open

Let’s take a look at the ASA NAT table.

```
ciscoasa(config)# show xlate
1 in use, 1 most used
Flags: D - DNS, e - extended, I - identity, i - dynamic, r - portmap,
       s - static, T - twice, N - net-to-net
TCP PAT from DMZ:10.10.11.2 80-80 to outside:105.255.1.200 80-80
    flags sr idle 0:00:17 timeout 0:00:00

ciscoasa(config)# show nat

Auto NAT Policies (Section 2)
1 (DMZ) to (outside) source static WEB_SERVER 105.255.1.200  service tcp www www
    translate_hits = 1, untranslate_hits = 0
ciscoasa(config)# show xlate
1 in use, 1 most used
Flags: D - DNS, e - extended, I - identity, i - dynamic, r - portmap,
       s - static, T - twice, N - net-to-net
TCP PAT from DMZ:10.10.11.2 80-80 to outside:105.255.1.200 80-80
    flags sr idle 0:03:36 timeout 0:00:00
```

Introduction to Firewall Architecture Design

These are different design topologies where we describe how a customer is connected (using BGP or a default route) to one or more ISPs.

Various ISP Connection Types

- **Single-homed: You** are connected to a single ISP using a single link.
- **Dual-homed**: You are connected to a single ISP using dual links.
- **Single multi-homed**: You are connected to two ISPs using a single link.
- **Dual multi-homed**: you are connected to two ISPs using dual links

Single-Homed Architecture

The single-homed design means you have a single connection to a single ISP. With this design, you don't need BGP since there is only one exit path in your network. You might as well just use a static default route that points to the ISP.

The advantage of a single-homed link is that it's cost-effective; the disadvantage is that you don't have any redundancy. Your link is a single point of failure, but so is using a single ISP

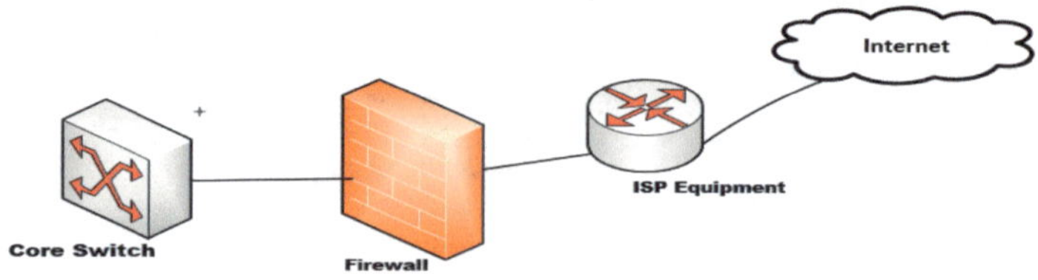

Single Multi-homed Architecture

Multihomed means we are connected to at least two different ISPs. The simplest design looks like this:

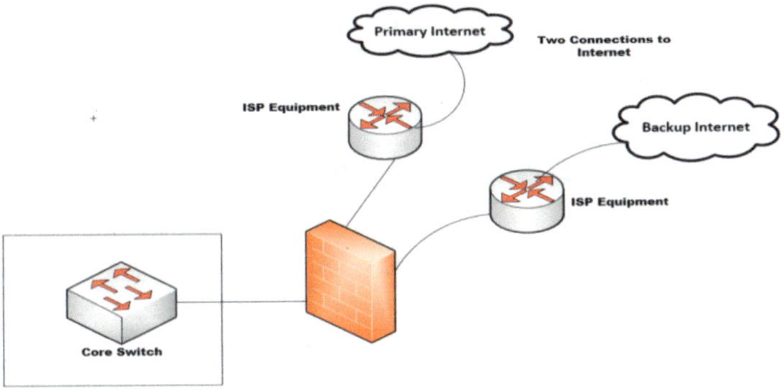

Above, you see that we have a single ASA or Router at the customer, connected to two different ISPs. The single point of failure in this design is that you only have one ASA at the customer. When it fails, you won't be able to connect to any ISP.

We can improve this by adding a second ASA, shown below. This is a pretty good design; we only use single links, but we are connected to two different ISPs using different routers/firewall.

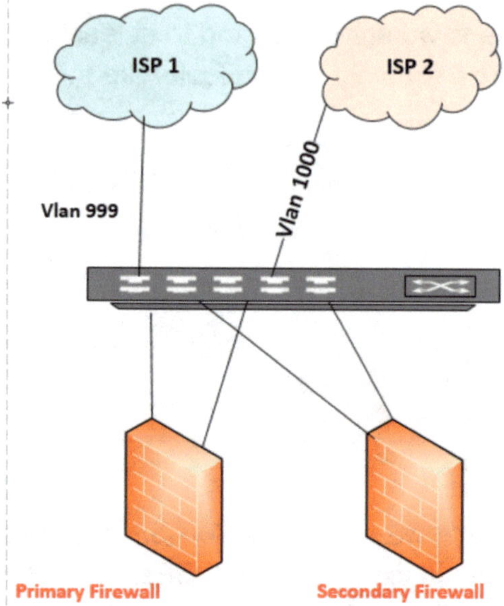
ISP 1
ISP 2
Vlan 999
Vlan 1000
Primary Firewall
Secondary Firewall

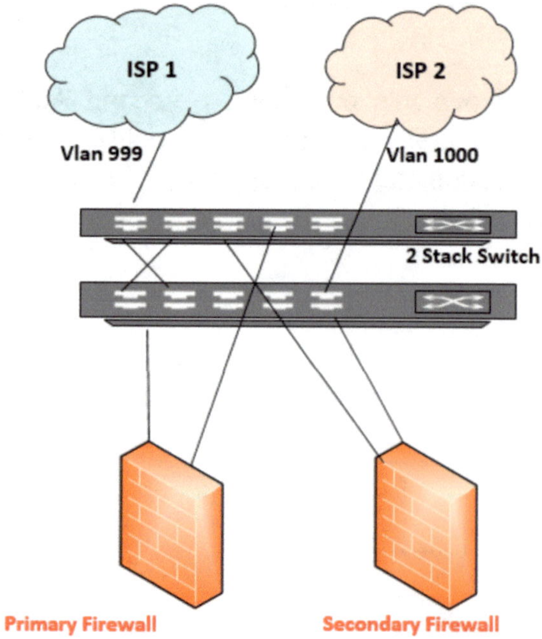
ISP 1
ISP 2
Vlan 999
Vlan 1000
2 Stack Switch
Primary Firewall
Secondary Firewall

Dual Multihomed

The dual multihomed design means we are connected to two different ISPs, and we use redundant links. There are some variations. Here's the first one:

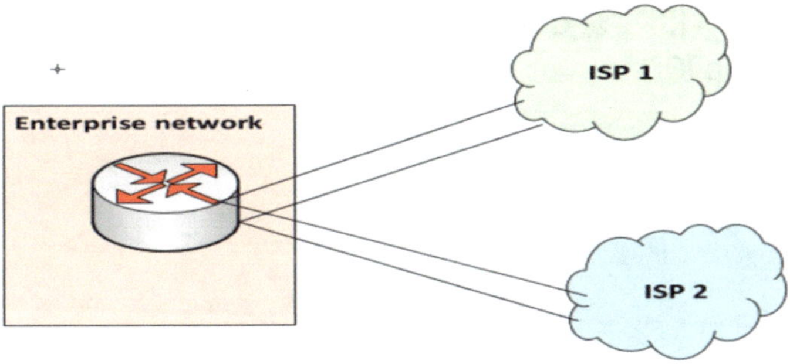

Using one router and two links to each ISP. We have redundant ISPs and links, but the router is still a single point of failure. We can improve this by adding a second router.

Dual WAN on Cisco ASA

Cisco ASA 5500 series firewall now supports the **Dual-ISP** capability. You can connect two interfaces of the firewall to two different ISPs and use the new "**SLA Monitor**" feature (SLA=Service Level Monitoring) to monitor the link to the primary ISP, and if that fails, the traffic is routed to the Backup ISP.

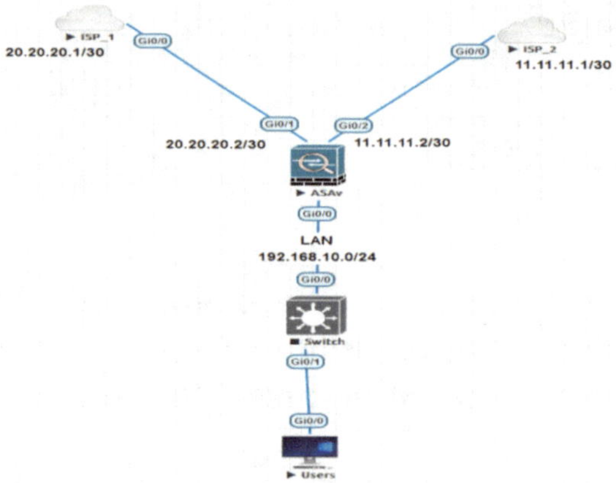

1. Configure all interfaces required, including the one to be use for the secondary ISP

ciscoasa(config-if)# interface GigabitEthernet0/0

ciscoasa(config-if)# nameif inside

ciscoasa(config-if)# security-level 100

ciscoasa(config-if)# ip address 192.168.10.1 255.255.255.0

ciscoasa(config-if)# no shut

ciscoasa(config-if)# interface GigabitEthernet0/1

ciscoasa(config-if)# nameif outside

ciscoasa(config-if)# security-level 0

ciscoasa(config-if)# ip address 20.20.20.2 255.255.255.252

ciscoasa(config-if)# no shut

ciscoasa(config-if)# interface GigabitEthernet0/2

cocoas(config-if)# nameif outside_backup00

ciscoasa(config-if)# security-level 0

ciscoasa(config-if)# ip address 11.11.11.2 255.255.255.252

cocoasa(config-if)# no shut

2. Configure the two ISP routers base on the topology and define a default route

Router(config)#hostname ISP

ISP-1(config)#int g0/0

ISP-1(config-if)#ip add 20.20.20 .1 255.255.255.252

ISP-1(config-if)#no shut

ISP-1(config-if)#int l0

ISP-1(config-if)#ip add 8.8.8.8 255.255.255.255

ISP-1(config-if)#ip route 0.0.0.0 0.0.0.0 20.20.20.2

ISP-1(config)#**do ping 20.20.20.2**

Type escape sequence to abort.

Sending 5, 100-byte ICMP Echos to 20.20.20.2, timeout is 2 seconds:

!!!!!

Success rate is 100 percent (5/5), round-trip min/avg/max = 3/4/8 ms

Router(config)#hostname ISP_2

ISP-2(config-if)#ip add 11.11.11.1 255.255.255.252

ISP-2(config-if)#no shut

ISP-2(config-if)#int l0

ISP-2(config-if)#ip add 8.8.8.8 255.255.255.255

ISP-2(config-if)#ip route 0.0.0.0 0.0.0.0 11.11.11.2

ISP-2(config)#do ping 11.11.11.2

Type escape sequence to abort.

Sending 5, 100-byte ICMP Echos to 11.11.11.2, timeout is 2 seconds:

!!!!!

Success rate is 100 percent (5/5), round-trip min/avg/max = 3/4/8 ms

USER1(config)# Hostname USER1

USER1(config)#int g0/0

USER1(config-if)#ip add 192.168.10.2 255.255.255.0

USER1(config-if)#ip route 0.0.0.0 0.0.0.0 192.168.10.1

USER1(config)#do ping 192.168.10.1

Type escape sequence to abort.

Sending 5, 100-byte ICMP Echos to 192.168.10.1, timeout is 2 seconds:

!!!!!

Success rate is 100 percent (5/5), round-trip min/avg/max = 10/12/15 ms

3. Enable icmp inspection on the cisco ASA to allow users to ping across the ASA

```
policy-map global_policy
class inspection_default
 inspect icmp
```

USER1(config)#do ping 20.20.20.1

Type escape sequence to abort.

Sending 5, 100-byte ICMP Echos to 20.20.20.1, timeout is 2 seconds:

!!!!!

USER1 (config)#do ping 11.11.11.1

Type escape sequence to abort.

Sending 5, 100-byte ICMP Echos to 11.11.11.1, timeout is 2 seconds:

!!!!!

Success rate is 100 percent (5/5), round-trip min/avg/max = 18/22/33

Let's test for 8.8.8.8, which is not supposed to work till we define the Active path.

USER1 (config)#do ping 8.8.8.8

Type escape sequence to abort.

Sending 5, 100-byte ICMP Echos to 8.8.8.8, timeout is 2 seconds:

.....

Success rate is 0 percent (0/5)

4. **Configure monitoring of the ISP's availability**

In order for the Cisco ASA firewall to monitor the availability of the primary channel, we need to configure the "ip sla monitor" function. It allows sending a ping request (an ICMP echo request) to ISP 1's gateway address at configured time intervals. Receiving a response (ICMP echo reply) will mean that the channel is available.

Additional information:

timeout 3000 – is the timeframe within which Cisco ASA will wait for an ICMP response. 3000 => 3 seconds
frequency 10 – how often to send the requests. Here is every 10 seconds

```
ciscoasa(config)# sla monitor 100
ciscoasa (config-sla-monitor)# type echo protocol
ipIcmpEcho 20.20.20.1 interface outside
ciscoasa (config-sla-monitor-echo)# timeout 3000
ciscoasa (config-sla-monitor-echo)# frequency 10
ciscoasa (config)# sla monitor schedule 100 life forever
start-time now
ciscoasa (config)# track 1 rtr 100 reachability
```

4. Configure the default gateway for the backup ISP

Just like with the primary ISP, the backup ISP needs to have its **default gateway** configured on the **Cisco ASA**, so that the firewall will send all the unknown packets in its direction. The only difference is that this gateway should **only** be used in case the **primary** ISP is unavailable, and not clog the routing table in all other cases. To achieve this, we need to change the **administrative distance** of the route – make it bigger, thus **lowering** the priority of this route. By default, all static routes have an administrative distance of **1**. We will configure an administrative distance of **254** for our **backup** channel, bringing it closer to the highest possible value.

ciscoasa(config-if)# route outside 0.0.0.0 0.0.0.0 20.20.20.1 1 track 1

ciscoasa(config)# route outside_backup 0.0.0.0 0.0.0.0 11.11.11.1 254

5. **Checking the IP SLA monitoring function state**

ciscoasa(config)# show sla monitor operational-state

Entry number: 100

Modification time: 21:14:48.683 UTC Mon Sep 23 2024

Number of Octets Used by this Entry: 1456

Number of operations attempted: 30

Number of operations skipped: 0

Current seconds left in Life: Forever

Operational state of entry: Active

Last time this entry was reset: Never

Connection loss occurred: FALSE

Timeout occurred: TRUE

Over thresholds occurred: FALSE

Latest RTT (milliseconds): No Connection/Busy/Timeout

Latest operation start time: 01:49:18.706 UTC Tue Sep 24 2024

Latest operation return code: Timeout

6. **Test to see if the users can access the internet**

USER1#ping 8.8.8.8

Type escape sequence to abort.

Sending 5, 100-byte ICMP Echos to 8.8.8.8, timeout is 2 seconds:

!!!!!

Success rate is 100 percent (5/5), round-trip min/avg/max = 15/20/29 ms

7. Do a ping test on a user and disable the primary connections, and see if the packet will drop

USER1#ping 8.8.8.8 re 1000

User1#ping 8.8.8.8

Type escape sequence to abort.

Sending 5, 100-byte ICMP Echos to 8.8.8.8, timeout is 2 seconds:

!!!!!

Success rate is 100 percent (5/5), round-trip min/avg/max = 20/22/25 ms

User1#ping 8.8.8.8 re 1000

Type escape sequence to abort.

Sending 1000, 100-byte ICMP Echos to 8.8.8.8, timeout is 2 seconds:

!!

!!

!!

!!!!!!!!!!!!!!!!!.....!!

!!

!!

Success rate is 99 percent (994/1000), round-trip min/avg/max = 6/18/55 ms

NB: when shut the on ISP we had just a few drop

Cisco ASA Policy-Based Routing (PBR) with Dual ISP

Policy-Based Routing (PBR) is a feature that has been supported on Cisco Routers for ages. However, Cisco ASA firewalls didn't support this until version 9.4.1 and later. **Policy-based routing** (**PBR**) is a technique used to make routing decisions based on policies set by the network administrator.

When a router receives a packet, it normally decides where to forward it based on the destination address in the packet, which is then used to look up an entry in a routing table. However, in some cases, there may be a need to forward the packet based on other criteria

PBR allows routing to be performed based on criteria other than the destination IP address. The traditional form of routing (which is used by default on any routing device) is based on the destination IP address of the packet.

With PBR, the network device can make routing decisions based on various other criteria such as source IP address, source port, protocol, destination port, etc., and also a combination of these.

This means, for example, that a routing device can receive a packet and look at its source IP address (instead of destination) and route the packet according to its PBR policy.

Many Enterprises utilize two ISP connections for redundancy and for bandwidth efficiency reasons.

One popular scenario, therefore, is to route some traffic to ISP1 and some other traffic to ISP2. For example, you can route all Web traffic (HTTP, HTTPS) through ISP1 and all other traffic through ISP2.

Another example could be to route traffic originating from the Engineering department via ISP1 and traffic originating from the Accounting department to go through ISP2.

Policy-Based Routing According to the Destination Protocol

Many Enterprises utilize two ISP connections for redundancy and for bandwidth efficiency reasons.

One popular scenario, therefore, is to route some traffic to ISP1 and some other traffic to ISP2. For example, you can route all Web traffic (HTTP, HTTPS) through ISP1 and all other traffic through ISP2.

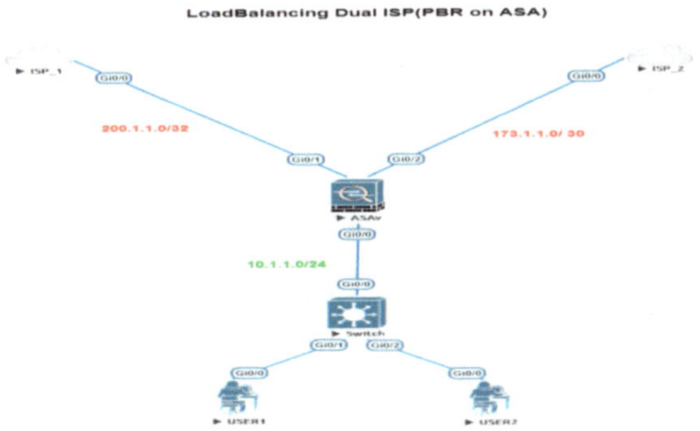

Project Task: The requirement is to route Web traffic (HTTP port 80 and HTTPS port 443) via ISP01 and all the other Internet traffic via ISP02.

Step1. First, configure the interfaces

```
ciscoasa(config)# int g0/0
ciscoasa(config-if)#  des link to LAN
ciscoasa(config-if)# nameif inside
INFO: Security level for "inside" set to 100 by default.
ciscoasa(config-if)# security-level 100
ciscoasa(config-if)# ip add 10.1.1.1 255.255.255.0
ciscoasa(config-if)# no shut
ciscoasa(config-if)# int g0/1
ciscoasa(config-if)# des link to ISP_1
INFO: Security level for "ISP01" set to 0 by default.
ciscoasa(config-if)# security-level 0
ciscoasa(config-if)# ip add 200.1.1.2 255.255.255.252
ciscoasa(config-if)# no shut
ciscoasa(config-if)# int g0/2
ciscoasa(config-if)# nameif ISP02
INFO: Security level for "ISP02" set to 0 by default.
ciscoasa(config-if)# security-level 0
ciscoasa(config-if)# ip add 173.1.1.2 255.255.255.252
ciscoasa(config-if)# no shut
```

Step2. Configure both ISPs with the following

Router(config)#hostname ISP_1

ISP_1(config)#

ISP_1(config)#int g0/0

ISP_1(config-if)#ip add 200.1.1.1 255.255.255.252

ISP_1(config-if)#no shut

ISP_1(config)#ip route 0.0.0.0 0.0.0.0 200.1.1.2

ISP_1(config)#ip http server

ISP_1(config)#ip http secure-server

ISP_1(config)#do ping 200.1.1.2

Type escape sequence to abort.

Sending 5, 100-byte ICMP Echos to 200.1.1.2, timeout is 2 seconds:

!!!!!

Success rate is 100 percent (5/5), round-trip min/avg/max = 3/8/23 ms

Router(config)#hostname ISP_2

ISP_2(config)#int g0/0

ISP_2(config-if)#ip add 173.1.1.1 255.255.255.252

ISP_2(config-if)#no shut

ISP_2(config)#ip route 0.0.0.0 0.0.0.0 173.1.1.2

ISP_2(config)#ip dns server

ISP_2(config-if)#line vty 0 4

ISP_2(config-line)#pass cisco

ISP_2(config-line)#transport input telnet

ISP_2(config-line)#login local

Router(config)#hostname USER1

USER1(config)#int g0/0

USER1(config-if)#ip add 10.1.1.2 255.255.255.0

USER1(config-if)#no shut

USER1(config-if)#ip route 0.0.0.0 0.0.0.0 10.1.1.1

USER1(config)#do ping 10.1.1.1

Sending 5, 100-byte ICMP Echos to 10.1.1.1, timeout is 2 seconds:

!!!!!

Success rate is 100 percent (5/5), round-trip min/avg/max = 8/12/18 ms

USER2(config)#int g0/0

USER2(config-if)#ip add 10.1.1.3 255.255.255.0

USER2(config-if)#no shut

USER2(config-if)#ip route 0.0.0.0 0.0.0.0 10.1.1.1

USER2(config)#do ping 10.1.1.1

Sending 5, 100-byte ICMP Echos to 10.1.1.1, timeout is 2 seconds:

!!!!!

Step 3: Define a default route on the ASA with HTTP and HTTPS traffic via ISP01 using an AD of 50 and all other traffic to ISP02 with an AD of 1.

Cocoasa (config-if) # route ISP01 0 0 200.1.1.1 50

Ciscoasa(config) # route ISP02 0 0 173.1.1.1 1

ciscoasa(config)# show route

The gateway of last resort is 173.1.1.1 to network 0.0.0.0

S* 0.0.0.0 0.0.0.0 [1/0] via 173.1.1.1, ISP02

C 10.1.1.0 255.255.255.0 is directly connected, inside

L 10.1.1.1 255.255.255.255 is directly connected, inside

C 173.1.1.0 255.255.255.252 is directly connected, ISP02

L 173.1.1.2 255.255.255.255 is directly connected, ISP02

C 200.1.1.0 255.255.255.252 is directly connected, ISP01

L 200.1.1.2 255.255.255.255 is directly connected, ISP01

Step 4: Configure NAT rules (PAT) using the corresponding outgoing interface of the ASA for traffic going from "inside" to "ISP01" and also for "inside" to "ISP02."

ciscoasa(config)# nat (inside, ISP01) 1 source dynamic any interface

ciscoasa(config)# nat (inside, ISP02) 2 source dynamic any interface

ciscoasa(config)# sho nat de

Manual NAT Policies (Section 1)

1 (inside) to (ISP01) source dynamic any interface

translate_hits = 0, untranslate_hits = 0

Source - Origin: 0.0.0.0/0, Translated: 200.1.1.2/30

2 (inside) to (ISP02) source dynamic any interface

translate_hits = 0, untranslate_hits = 0

Source - Origin: 0.0.0.0/0, Translated: 173.1.1.2/30

Step 5: Create an Access Control List (ACL) that will match the traffic that we want to be handled by our PBR policy.

Create an object group for ports 80,443

ciscoasa(config)# object-group service WEB-ports tcp

ciscoasa(config-service-object-group)# port-object eq 443

ciscoasa(config-service-object-group)# port-object eq 80

ciscoasa(config)# access-list PBR_ACL extended permit tcp any any object-group WEB-ports

The ACL above matches traffic from any inside network having destination ports of 80 and 443

Step 6: We need to create a route-map that will match the traffic in the ACL created above, and then apply a routing policy to this traffic flow.

ciscoasa(config)# route-map PBR permit 2

match ip address PBR_ACL (**match the traffic identified in the ACL created above)**

ciscoasa(config-route-map)# set ip next-hop 200.1.1.1 (set the next hop of the traffic to be ISP01)

ciscoasa(config)# show route-map

route-map PBR, permit, sequence 2

Match clauses:

ip address (access-lists): PBR_ACL

Set clauses:

ip next-hop 200.1.1.1

Step 7. Apply the PBR policy to the "Ingress" interface that we want to enforce this routing policy

ciscoasa(config-route-map)# int g0/0

ciscoasa(config-if)# policy-route route-map PBR (**apply the PBR policy to this interface)**

ciscoasa(config-if)# show run int g0/0

interface GigabitEthernet0/0

nameif inside

security-level 100

ip address 10.1.1.1 255.255.255.0

policy-route route-map PBR

Let's generate traffic to test the connection

USER!#**telnet 8.8.8.8 443**

Trying 8.8.8.8, 443 ... Open

USER!#**telnet 8.8.8.8 80**

Trying 8.8.8.8, 80 ... Open

Test Connection using Packet-TRACER

ciscoasa(config-if)# packet-tracer input inside tcp 10.1.1.2 1 8.8.8.8 443

Result:

input-interface: inside

input-status: up

input-line-status: up

output-interface: ISP01

output-status: up

output-line-status: up

Action: allow

Policy-Based Routing According to Source Networks

Another example could be to route traffic originating from the Engineering department via ISP1 and traffic originating from the Accounting department to go through ISP2.

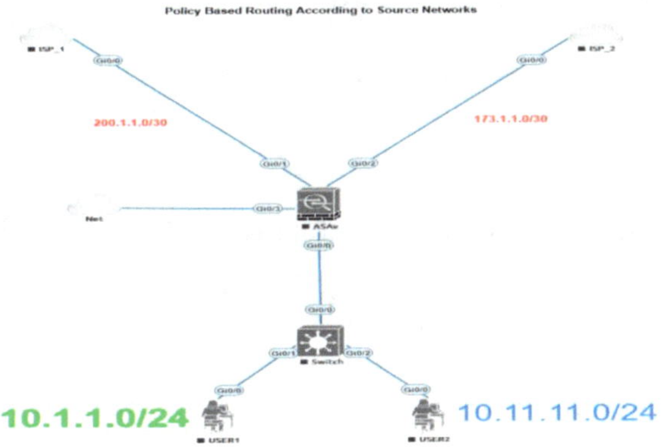

Project Task: The requirement is to route traffic originating from the Engineering department via ISP1 and traffic originating from the accounting department to go through ISP2

Step1.Create the interface for LAN1 and LAN2

ciscoasa(config)# interface GigabitEthernet0/0

ciscoasa(config-if)# no nameif

ciscoasa(config-if)# no security-level

ciscoasa(config-if)# no ip address

ciscoasa(config-if)# do show run no shut

ciscoasa(config-if)# interface GigabitEthernet0/0.100

ciscoasa(config-subif)# vlan 100

ciscoasa(config-subif)# nameif LAN1

ciscoasa(config-subif)# security-level 100

ciscoasa(config-subif)# ip address 10.1.1.1 255.255.255.0

ciscoasa(config-subif)# interface GigabitEthernet0/0.200

ciscoasa(config-subif)# vlan 200

ciscoasa(config-subif)# nameif LAN2

ciscoasa(config-subif)# security-level 100

ciscoasa(config-subif)# ip address 10.11.11.1
255.255.255.0

Step 2: Create an Access Control List (ACL) that will match the traffic that we want to be handled by our PBR policy.

ciscoasa(config)#access-list PBR_ACL1 extended permit ip 10.1.1.0 255.255.255.0 any

ciscoasa(config)#access-list PBR_ACL2 extended permit ip 10.11.11.0 255.255.255.0 any

Step-3 We'll configure the route-map, which will match the traffic in ACLs created above, and then apply a routing policy to the traffic flows.

route-map PBR permit 2 **create the route-map and give it a name "PBR"**

match ip address PBR_ACL1**<– match the traffic of LAN1 identified in ACL1 created**

set ip next-hop 200.1.1.1**<– set the next hop of LAN1 traffic to be ISP1**

route-map PBR permit 3**<– create another entry in the same route-map**

match ip address PBR_ACL2**<– match the traffic of LAN2 identified in ACL2 created**

set ip next-hop 173.1.1.1**<– set the next hop of LAN2 traffic to be ISP2**

Step 4: Apply the PBR policy to the "Ingress" interfaces that we want to enforce this routing policy. In our case, we will apply the same policy to both internal networks (LAN1, LAN2)

interface GigabitEthernet0/0.100

policy-route route-map PBR

interface GigabitEthernet0/0.200

policy-route route-map PBR

Step 5: Again, we need to take care of NAT since we must translate the private internal IP networks to a public IP address in order to access the Internet.

nat (LAN1, ISP01) source dynamic any interface

nat (LAN2, ISP02) source dynamic any interface

lets verify

packet-tracer input LAN2 tcp 10.11.11.3 80 8.8.8.8 80

Phase: 1

Type: PBR-LOOKUP

Subtype: policy-route

Result: ALLOW

Config:

route-map PBR permit 3

match ip address PBR_ACL2

set ip next-hop 173.1.1.1

Additional Information:

Matched route-map PBR, sequence 3, permit

Found next-hop 173.1.1.1 using egress ifc ISP02

Phase: 2

Type: NAT

Subtype:

Result: ALLOW

Config:

nat (LAN2,ISP02) source dynamic any interface

Additional Information:

Dynamic translate 10.11.11.3/80 to 173.1.1.2/8

Result:

output-interface: ISP02

output-status: up

output-line-status: up

Action: allow

Ciscoasa (config)# packet-tracer input LAN1 tcp 10.1.1.3 1 8.8.8.8 80

Phase: 1

Type: PBR-LOOKUP

Subtype: policy-route

Result: ALLOW

Config:

route-map PBR permit 2

match ip address PBR_ACL1

set ip next-hop 200.1.1.1

Additional Information:

Matched route-map PBR, sequence 2, permit

Found next-hop 200.1.1.1 using egress ifc ISP01

Phase: 2

Type: NAT

Subtype:

Result: ALLOW

Config:

nat (LAN1, ISP01) source dynamic any interface

Additional Information:

Dynamic translate 10.1.1.3/1 to 200.1.1.2/1

Result:

output-interface: ISP01

output-status: up

output-line-status: up

Action: allow

Firewall Redundancy

Using just a single ASA is a single point of failure and usually catastrophically reflects in the network when the device experiences common setbacks such as hardware issues, link/cable problems, or just a simple misconfiguration.

Therefore, using a second ASA to the primary one will provide a backup solution in case something goes wrong with the active unit

Overall, the deployment of multiple firewalls offers a variety of benefits, ranging from greater performance to enhanced security. If your security environment warrants this type of scenario, it's definitely an option worth considering.

There are 3 common Firewall redundancy designs generally practiced in the industry.

Common Deployment Scenarios

1. Fault tolerance and load balancing
2. Enhanced perimeter protection
3. Protected subnets, Redundancy firewall Design

Deployment Scenarios and Benefits

Fault Tolerance and Load Balancing, Redundancy, Firewall Design

Many organizations choose to implement dual firewalls in a parallel fashion, as shown in the figure below. When the router is properly configured, this provides the added benefits of fault tolerance and load balancing. Both firewalls should be configured to "fail-safe," that is, in the event of a failure, they should automatically block all traffic. When configured in this fashion, the firewalls provide fault tolerance; when one fails, the other is able to carry the network traffic and keep the failure transparent to users.

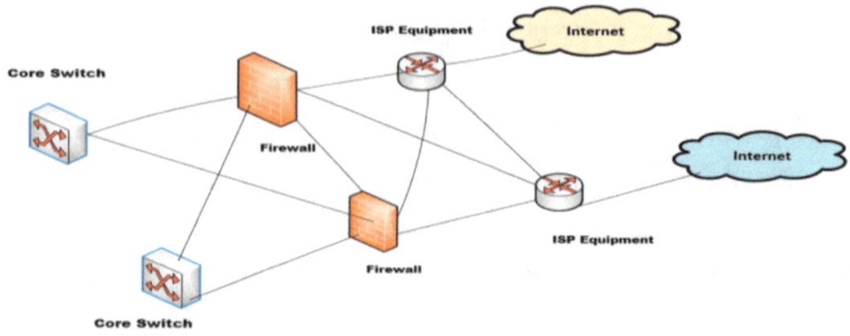

Enhanced Perimeter Protection, Redundancy, Firewall Design

It's also possible to deploy the two firewalls in a series circuit, as shown in the illustration below. When configured in this fashion, all traffic passing into or out of the network must pass through both firewalls. This setup is sometimes deployed in high-security environments to protect against firewall-specific vulnerabilities. In this case, the two firewalls are from different vendors and may even run on different operating systems.

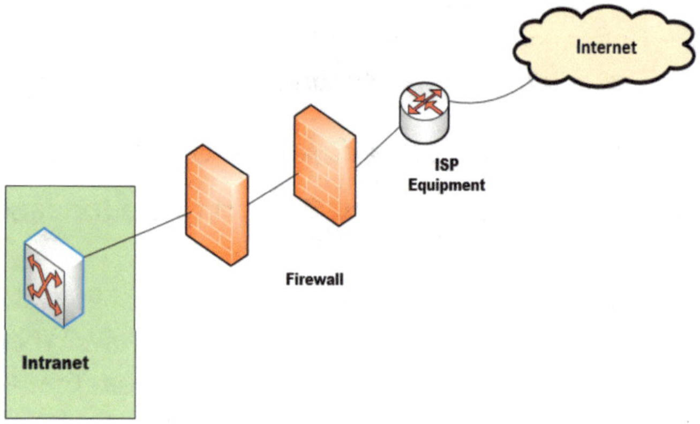

Protected subnets, Redundancy firewall Design

The final scenario we'll discuss is shown in the figure below. In this case, secondary firewall(s) are used to protect subnets of the internal network that have greater security requirements than the network as a whole. This type of scenario may be used, for example, to provide an accounting department with added protection for sensitive financial data they wish to protect from other internal users.

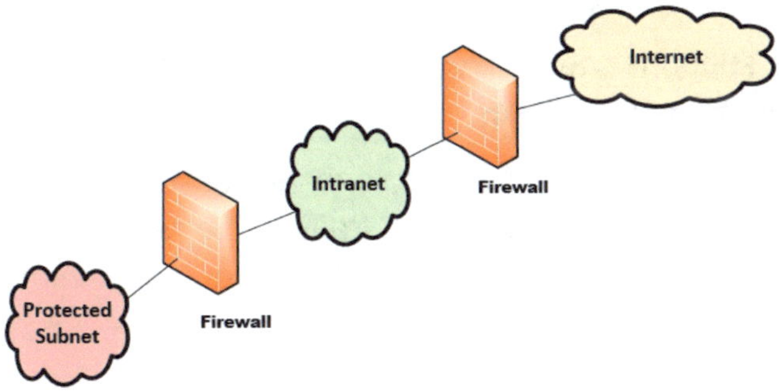

ASA Failover

ASA failover refers to the capability of Cisco Adaptive Security Appliances (ASAs) to automatically switch to a backup unit in the event of a primary unit failure. It creates a seamless transition, maintaining network connectivity without any noticeable interruption. ASA failover operates in Active/Standby and Active/Active modes.

Cisco ASA Failover Modes

ASA supports two failover modes, Active/Active failover and Active/Standby failover.

In **Active/Standby** failover, one device functions as the **Active Unit** and passes the traffic. The second **Standby Unit** does not actively pass traffic. When a failover occurs, the Standby unit assumes the active role and starts passing the traffic.

In an **Active/Active** failover, **both** ASAs can pass traffic. Please note that Active/Active failover is **only** available to ASAs in **multiple context models**. In Active/Active failover, you divide the security contexts on the ASA into 2 failover groups. A failover group is simply a logical group of one or more security contexts. One group is assigned to be Active on the primary ASA, and the other group is assigned to be active on the Secondary ASA. When a failover occurs, it occurs at the failover group level.

Failover Types

Within these two different failover modes, there are also two different failover types: stateless and stateful.

When using **stateless failover,** if a failover should need to occur, all active connections will be dropped and will have to be reestablished to continue communications.

When using **stateful failover**, connection state information is exchanged between the failover partners (or groups). If a failover should need to occur, the active connections (that are supported) can be seamlessly transferred and will not need to be reestablished.

Failover Triggers

Failover can be triggered at the unit level if one of the following events occurs:

- The unit has a hardware failure.
- The unit has a power failure.
- The unit has a software failure.

ASA Failover Requirement

If you want to use failover, you must meet the following requirements:

Hardware:

- ASA failover platform must be the same model. For example, 2x ASA 5510 or 2x ASA 5522.
- ASA failover platform must have the same number and types of interfaces.
- ASA failover platform must have the same modules installed (if any are to be installed).
- ASA failover platform must have the same amount of RAM installed (it is also preferred if the Flash sizes are the same as well).

Software:

- Both ASA failover platforms must be using the same firewall mode (routed or transparent).
- Both ASA failover platforms must be using the same context mode (single or multiple).
- Both ASA failover platforms must be using the same major and minor software version (there are exceptions during upgrade).

- Both ASA failover platforms must use the same AnyConnect images.

The failover mechanism is stateful, which means that the active ASA sends all stateful connection information to the standby ASA. This includes TCP/UDP states, NAT translation tables, ARP table, VPN information and more.

Benefits of Cisco ASA Failover

ASA Failover offers numerous benefits for businesses.

Enhanced Network Uptime: Organizations can achieve uninterrupted network connectivity with Cisco ASA failover. In the event of a primary unit failure, the secondary unit seamlessly takes over, ensuring minimal disruption to network operations.

ASA Failover enhances security by providing seamless failover for security policies, preventing potential vulnerabilities during critical moments

Load Balancing: Cisco ASA failover enables load balancing, distributing incoming network traffic across multiple units. This optimizes resource utilization and prevents any single unit from becoming overloaded.

Improved Scalability: Failover setup allows for easy scalability, as additional units can be added to the configuration. This helps accommodate growing network demands without compromising on security or performance.

Active/Standby ASA Firewall Project

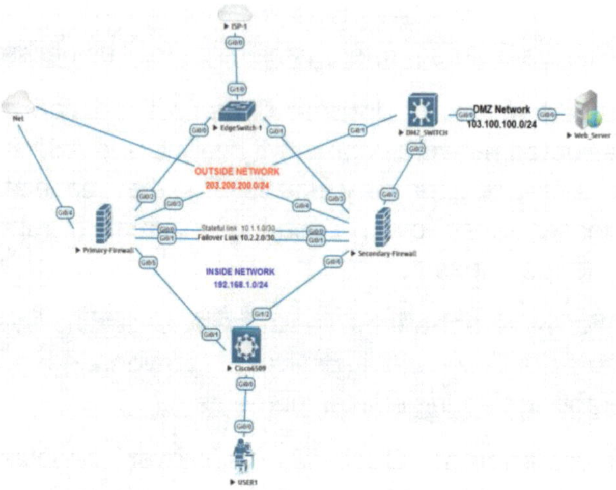

ASA Failover Configuration Guide

Active Unit Configuration:

Note: Always start with the active ASA first.

Step 1: Assign an IP address to the outside interface. During Failover, the primary IP address will be assigned to the Standby Unit

PrimaryFW(config-if)# int g0/2

PrimaryFW(config-if)# des to ISP

PrimaryFW(config-if)# nameif outside

PrimaryFW(config-if)# security-level 0

PrimaryFW(config-if)# ip add 203.200.200.1 255.255.255.0 standby 203.200.200.2

Step 2: Assign an IP address to the inside interface. During Failover, the primary IP address will be assigned to the Standby Unit

PrimaryFW(config-if)# int g0/5

PrimaryFW(config-if)# des to CoreSwitch

PrimaryFW(config-if)# nameif inside

PrimaryFW(config-if)# security-level 100

PrimaryFW(config-if)# ip add 192.168.1.1 255.255.255.0 standby 192.168.1.2

Step 3: Assign an IP address to the DMZ interface. During Failover, the primary IP address will be assigned to the Standby Unit

PrimaryFW(config-if)# int g0/3

des link to DMZ

nameif DMZ

ip add 103.100.100.1 255.255.255.0 standby 103.100.100.2

no shut

Step-4: Configure the routing protocol to allow users to communicate with the ISP

route outside 0.0.0.0 0.0.0.0 203.200.200.3

router eigrp 10

network 192.168.1.0 255.255.255.0

network 103.100.100.0 255.255.255.0

redistribute static

USER#ping 8.8.8.8

Type escape sequence to abort.

Sending 5, 100-byte ICMP Echos to 8.8.8.8, timeout is 2 seconds:

!!!!!

Success rate is 100 percent (5/5), round-trip min/avg/max = 15/17/19 ms

USER#ping 103.100.100.3

Type escape sequence to abort.

Sending 5, 100-byte ICMP Echos to 103.100.100.3, timeout is 2 seconds:

!!!!!

Success rate is 100 percent (5/5), round-trip min/avg/max = 13/15/21 ms

Step 5: Set ASA1 as the primary unit

PrimaryFW(config-if)# failover lan unit primary

Step-6 Define Failover Interface.

NB: used to determine the operating status of each unit, as well as to replicate and synchronize any configuration between both units in the pair.

In order to accomplish this, a dedicated Ethernet interface must be used on each Cisco ASA, which will be used exclusively for passing failover information. The connection between these interfaces on the ASAs can either be a direct link or through a switch.

PrimaryFW(config-if)# failover lan interface FAILOVER-LINK G0/1

INFO: Non-failover interface config is cleared on GigabitEthernet0/1 and its sub-interfaces

Step 7: Assign IP address to Failover Interfaces

PrimaryFW(config-if)# failover interface ip FAILOVER-LINK 10.2.2.1 255.255.255.252 standby 10.2.2.2

Step 8: Define a stateful Failover interface

NB: Without this data, anytime a failover happens, all end-user applications must re-establish connections, and there will be an interruption from the client's point of view. The stateful failover interface can either be a dedicated interface or shared with any other interface, including the LAN failover interface.

Some of the information exchanged over the stateful failover link is:

Network Address Translation (NAT) table

Address Resolution Protocol (ARP) table

TCP connection table

UDP connection table

HTTP connection table

MAC address table

PrimaryFW(config-if)# failover link Stateful-link G0/0

INFO: Non-failover interface config is cleared on GigabitEthernet0/0 and its sub-interfaces

Step 9: Assign IP addresses to Stateful Failover interfaces

PrimaryFW(config-if)# failover interface ip stateful-link 10.1.1.1 255.255.255.252 standby 10.1.1.2

Step-10 Enable Failover

PrimaryFW(config-if)# failover

Note: Issue the failover command on the primary device first, and then issue it on the secondary device. After you issue the failover command on the secondary device, the secondary device immediately pulls the configuration from the primary device and sets itself as standby.

The primary ASA stays up and passes traffic normally and marks itself as the active device. From that point on, whenever a failure occurs on the active device, the standby device comes up as active.

Step 11: Enable the interfaces for Failover and State Link

PrimaryFW(config-if)# int g0/0

PrimaryFW(config-if)# no shut

PrimaryFW(config-if)# int g0/1

Let's verify what is configured

PrimaryFW(config-if)# show run failover

failover

failover lan unit primary

failover lan interface FAILOVER-LINK GigabitEthernet0/1

failover link Stateful-link GigabitEthernet0/0

failover interface ip FAILOVER-LINK 10.2.2.1 255.255.255.252 standby 10.2.2.2

failover interface ip Stateful-link 10.1.1.1 255.255.255.252 standby 10.1.1.2

The ASA requires something that triggers the failover mechanism. An interface that fails is a good trigger. When the inside or outside interface fails, we should fail over. By default, all physical interfaces are monitored, but let me show you the command anyway:

PrimaryFW(config)# monitor-interface inside

PrimaryFW(config)# monitor-interface outside

PrimaryFW(config)# monitor-interface DMZ

Configuration on Secondary ASA

Step-1 Set ASA1 as the primary unit

SecondaryFW(config-if)# failover lan unit Secondary

Step-2 Define Failover Interface.

SecondaryFW(config-if)# failover lan interfaceFAILOVER-LINK G0/1

INFO: Non-failover interface config is cleared on GigabitEthernet0/1 and its

Step 3: Assign IP address to Failover Interfaces

SecondaryFW(config-if)# failover interface ip FAILOVER 10.2.2.1 255.255.255.252 standby 10.2.2.2

Step-4 Enable the interfaces for Failover and State Link

SecondaryFW(config-if)# int g0/0

SecondaryFW(config-if)# no shut

SecondaryFW(config-if)# int g0/1

SecondaryFW(config-if)# no shut

Step-5 Enable Failover

SecondaryFW(config-if)# failover

Step-6 Change the prompt to show primary or secondary

Once the configuration is replicated on both ASAs, they both use the same hostname. This means that every time you access the ASAs over a console, Secure Shell (SSH), or

Telnet connection, it will not be easy to differentiate between the units being managed.

Therefore, it is recommended to change the CLI prompt to include additional information next to the hostname, such as the priority and state of the managed device.

PrimaryFW(config-if)# prompt hostname priority state

PrimaryFW/pri/act(config)#

PrimaryFW/pri/act(config)#

Now let's check on Secondary FW

PrimaryFW/sec/stby>

PrimaryFW/sec/stby>

This is what you will see on Primary ASA1

PrimaryFW(config)#

Switchover enabled

Configuration has changed, replicate to mate.

Beginning configuration replication: Sending to mate.

End Configuration Replication to mate

Switching to Standby

Primary: Switching to Ok for reason Interface check.

Switching to Active

This is what you will see on Secondary ASA2

Switchover enabled

Configuration has changed, replicate to mate.

State check detected an Active mate

Beginning configuration replication from the mate.

End configuration replication from the mate.

PrimaryFW(config)# show failover

PrimaryFW/pri/act(config)# show failover

Failover On

Failover unit Primary

Failover LAN Interface: FAILOVER-LINK GigabitEthernet0/1 (up)

Reconnect timeout 0:00:00

Unit Poll frequency 1 seconds, holdtime 15 seconds

Interface Poll frequency 5 seconds, holdtime 25 seconds

Interface Policy 1

Monitored Interfaces 4 of 61 maximum

MAC Address Move Notification Interval not set

Version: Ours 9.6(1), Mate 9.6(1)

Serial Number: Ours 9AQFS5U27GJ, Mate 9AKXX77E3FA

Last Failover at: 22:40:12 UTC Oct 7 2024

This host: Primary - Active

Active time: 5576 (sec)

slot 0: empty

Interface outside (203.200.200.1): Normal (Monitored)

Interface dmz (103.100.100.1): Normal (Monitored)

Interface Management (10.255.1.201): Normal (Monitored)

Interface inside (192.168.1.1): Normal (Monitored)

Other host: Secondary - Standby Ready

Active time: 0 (sec)

Interface outside (203.200.200.2): Normal (Monitored)

Interface dmz (103.100.100.2): Normal (Monitored)

Interface Management (10.255.1.202): Normal (Monitored)

PrimaryFW/pri/act(config)# show monitor-interface

This host: Primary - Active

Interface outside (203.200.200.1): Normal (Monitored)

Interface dmz (103.100.100.1): Normal (Monitored)

Interface Management (10.255.1.201): Normal (Monitored)

Interface inside (192.168.1.1): Normal (Monitored)

Other host: Secondary - Standby Ready

Interface outside (203.200.200.2): Normal (Monitored)

Interface dmz (103.100.100.2): Normal (Monitored)

Interface Management (10.255.1.202): Normal (Monitored)

Interface inside (192.168.1.2): Normal (Monitored)

Verification

Ping the internet and power off the Primary FW

Router#ping 8.8.8.8 re 1000

Type escape sequence to abort.

Sending 1000, 100-byte ICMP Echos to 8.8.8.8, timeout is 2 seconds:

!!

*Oct 8 00:21:39.809: %DUAL-5-NBRCHANGE: EIGRP-IPv4 10: Neighbor 192.168.1.1 (GigabitEthernet0/0) is down: peer restarted...!!!!!!!!!!!!!!!!!!!!!!!!!!!!!!!!!!!

*Oct 8 00:21:44.588: %DUAL-5-NBRCHANGE: EIGRP-IPv4 10: Neighbor 192.168.1.1 (GigabitEthernet0/0) is up: new adjacency!!!!!

!!

!!

Failover exec

To execute a command on a specific unit in a failover pair, use the **failover exec** command in privileged EXEC or global configuration mode.

PrimaryFW/pri/act(config)# failover exec standby show failover

Failover On
Failover unit Secondary

Failover LAN Interface: FAILOVER_LINK GigabitEthernet0/1 (up)
Reconnect timeout 0:00:00
Unit Poll frequency 1 seconds, holdtime 15 seconds
Interface Poll frequency 5 seconds, holdtime 25 seconds
Interface Policy 1
Monitored Interfaces 4 of 61 maximum
MAC Address Move Notification Interval not set
Failover replication http
Version: Ours 9.6(1), Mate 9.6(1)
Serial Number: Ours 9AKXX77E3FA, Mate 9AQFS5U27GJ
Last Failover at: 01:21:45 UTC Oct 10 2024
This host: Secondary - Standby Ready
Active time: 0 (sec)
slot 0: empty
Interface outside (203.200.200.2): Normal (Monitored)
Interface DMZ (103.100.100.2): Normal (Monitored)
Interface Management (10.255.1.202): Normal (Monitored)
Interface inside (192.168.1.2): Normal (Monitored)

Other host: Primary - Active
Active time: 225420 (sec)
Interface outside (203.200.200.1): Normal (Monitored)

PrimaryFW/pri/act(config)#**failover exec mate show running-config failover**

failover
failover lan unit secondary
failover lan interface FAILOVER_LINK GigabitEthernet0/1
failover replication http
failover link Stateful-link GigabitEthernet0/0
failover interface ip FAILOVER_LINK 10.2.2.1 255.255.255.252 standby 10.2.2.2
failover interface ip Stateful-link 10.1.1.1 255.255.255.252 standby 10.1.1.2

PrimaryFW/pri/act(config)#**failover exec standby show interface**

PrimaryFW/pri/act(config)#**failover reload-standby**

Using ASDM to Configure ASA Failover

Step 1: Set up standby ip address

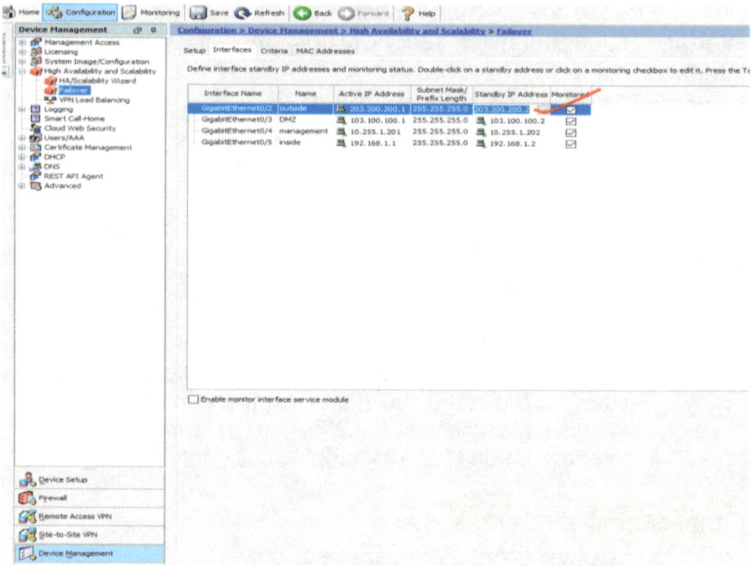

Step 2: Configure failover for Primary ASA

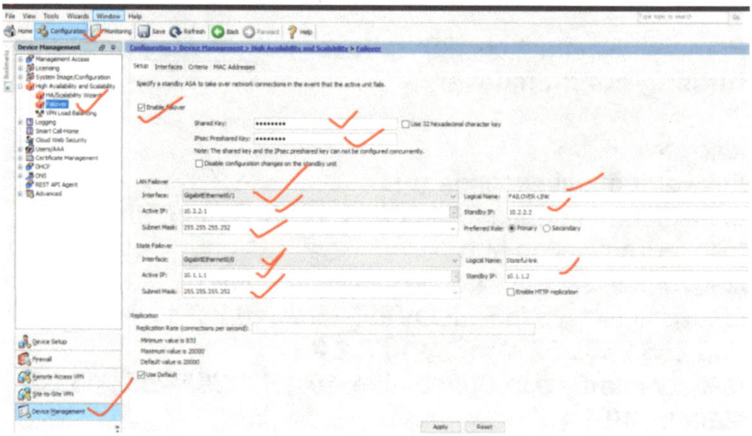

Step 3: Very your config

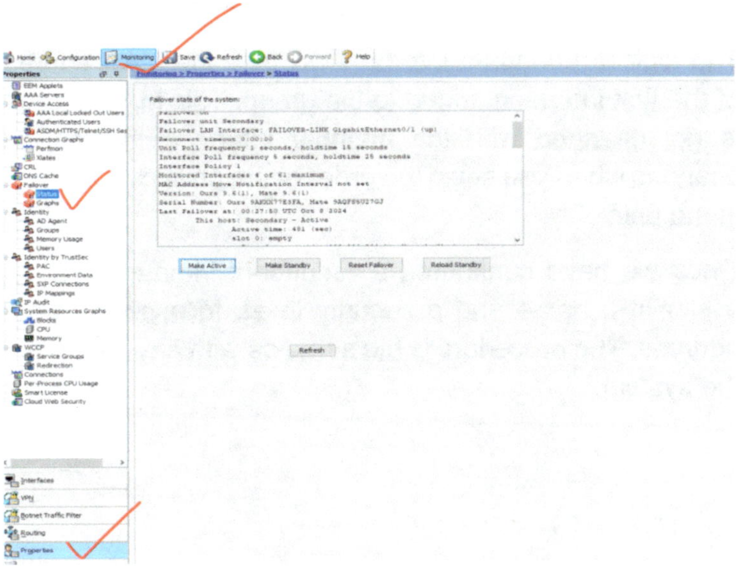

ASA Redundant Interface

A logical redundant interface consists of a pair of physical interfaces: an active and a standby interface. When the active interface fails, the standby interface becomes active and starts passing traffic. You can configure a redundant interface to increase the ASA reliability. This feature is separate from device-level failover, but you can configure redundant interfaces as well as device-level failover if desired.

The logical redundant interface is only available on ASA platforms, not on devices running Firepower.

We can configure up to 8 redundant interfaces.

Redundant interfaces are number from 1 to 8 and have the name redundant X. When adding physical interfaces to the redundant pair, please make sure there is no configuration on it and the interface is also in no shutdown state. This is just a precaution; the firewall will remove these settings when adding the physical interface to a new group.

The logical redundant interface will take the MAC address of the first interface added to the group. This MAC address is not changed with the member interface failures, but changes when you swap the order of the physical interfaces in the pair.

Once we have configured a redundant interface, we can assign it a name and a security level, followed by an IP address. The procedure is the same as with any interface in the system.

ASA Redundant Interface Project

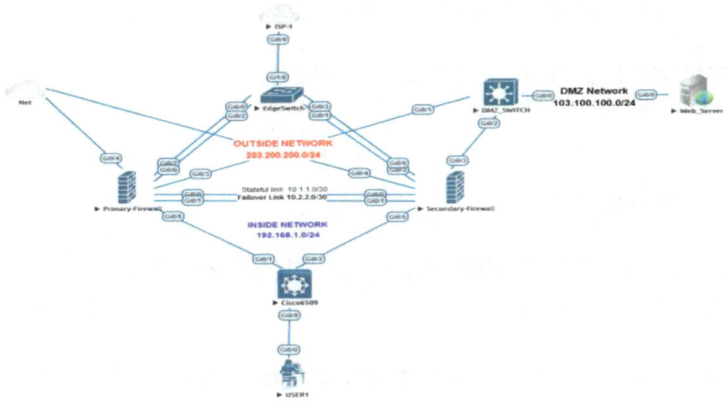

Using Command Line

interface GigabitEthernet0/2

no nameif

no security-level

No IP address

interface GigabitEthernet0/6

no nameif

no security-level

No IP address

Interface Redundant1

member-interface GigabitEthernet0/2

member-interface GigabitEthernet0/6

nameif outside

security-level 0

ip address 203.200.200.1 255.255.255.0 standby 203.200.200.2

Using ASDM to configure a redundant interface

Configure> device setup> interfaces>add

NB: Please make sure there is no configuration on it, and the interface is also in no shutdown state

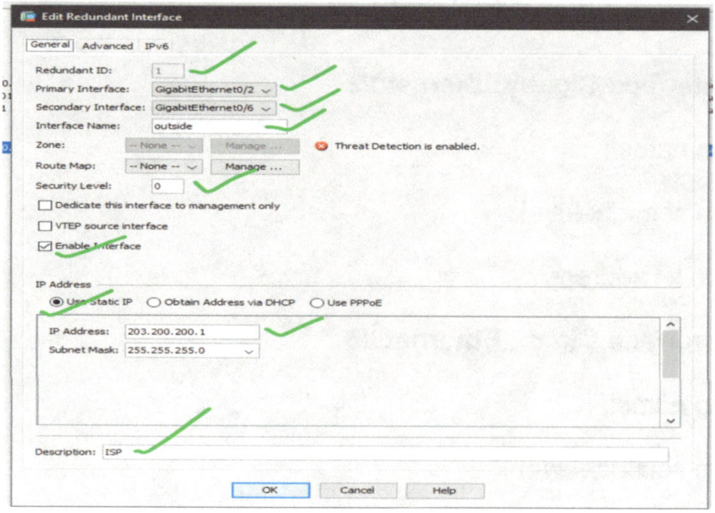

Verify

PrimaryFW/pri/act(config)# show run int r1

interface Redundant1

description ISP

member-interface GigabitEthernet0/2

member-interface GigabitEthernet0/6

nameif outside

security-level 0

ip address 203.200.200.1 255.255.255.0 standby 203.200.200.2

PrimaryFW/pri/act(config)# show interface redundant1

Interface Redundant1 "outside" is up, line protocol is up

Hardware is i82540EM rev03, BW 1000 Mbps, DLY 10 usec

Auto-Duplex(Full-duplex), Auto-Speed(1000 Mbps)

Input flow control is unsupported, output flow control is off

Description: ISP

Redundancy Information:

Member GigabitEthernet0/2(Active), GigabitEthernet0/6

Last switchover at 20:37:17 UTC Oct 14 2024

You can also force failover on the interface.

PrimaryFW/pri/act(config)# redundant-interface redundant 1 active-member GigabitEthernet0/6

PrimaryFW/pri/act(config)# show int r1

Redundancy Information:

Member GigabitEthernet0/6(Active), GigabitEthernet0/2

Last switchover at 22:59:28 UTC Oct 14 2024

Cisco ASA BotNet Filtering

Cisco ASA Adaptive Security Appliance is a Cisco proprietary firewall appliance device. ASA offers features like inspection, policing and prioritising traffic, filtering packets based on ACLs and Anti-X protection. The Anti-X features enable us to configure the botnet attack filter in Cisco ASA.

Botnet Filtering

The Cisco ASA Botnet Traffic Filter is an effective tool that enterprises can use to gain insights into one of today's leading threats. In conjunction with accurate threat data provided by Cisco Security Intelligence Operations and Cisco Global Correlation for IPS, the Botnet Traffic Filter offers an industry-leading solution to combat modern botnet threats in a dynamic business environment. These lists are stored in a database as per their reputations. Cisco ASA accesses the database, performing reputation-based filtering to identify the hacker.

Traffic classifications:

Traffic that passes through the Botnet Traffic Filter is classified into four categories:

Blacklist:

This is traffic to or from an IP address that is considered to be malicious. This IP address can be either an IP address/network entry in the dynamic blacklist or administrator-configured blacklist or it can be a snooped IP address that was found in a DNS reply for a blacklisted domain.

Whitelist:

This is traffic to or from an IP address that is considered to be good. It is part of the administrator-configured lists.

Greylist:

These addresses are associated with multiple domain names, but not all of these domain names are on the blacklist.

Unknown/None:

An IP address that does not map to a domain in either a blacklist or whitelist, and no syslog's will be generated for this traffic.

NB: Unlisted addresses do not generate any syslog messages, but addresses on the blacklist, whitelist, and greylist generate syslog messages.

Enabling the Adaptive Security Appliance to use Botnet Filtering requires a certain set of processes. Please remember, this feature works only with a license. The Cisco ASA appliance with the Botnet Traffic Filter should be deployed at the edge of the enterprise, as the botnet database contains information only about external botnets. It is also best to address the external threat as close to the source as possible. This feature is restricted to IPv4 traffic. The Botnet Traffic Filter is supported in all firewall modes (single and multiple) and in routed and transparent modes.

Requirement for Botnet Filtering

Cisco ASA Firewall must have a valid Botnet Filtering license and have access to Cisco's Security Intelligence Operation (CSIO) dynamic database, which is on the internet. This is essential as Botnet Filtering features would communicate with the CSIO dynamic database and verify with its White & Black listed databases.

DNS is required on ASA primarily for two reasons

To make sure the ASA is capable of resolving the Cisco Security Intelligence Operations server IP

Allow to have a static whitelist site, even if the site is blacklisted.

Configuration Steps

1. Enable DNS on the ASA
2. Configure a dynamic database
3. Configure the static database
4. Enable DNS snooping
5. Enable the Botnet traffic filter

Step 1: Configure the dynamic database

Via ASDM, the database can be configured through:

Step 1: Enable DNS on the ASA

Configuration ‣ Device Management‣ DNS‣ DNS client

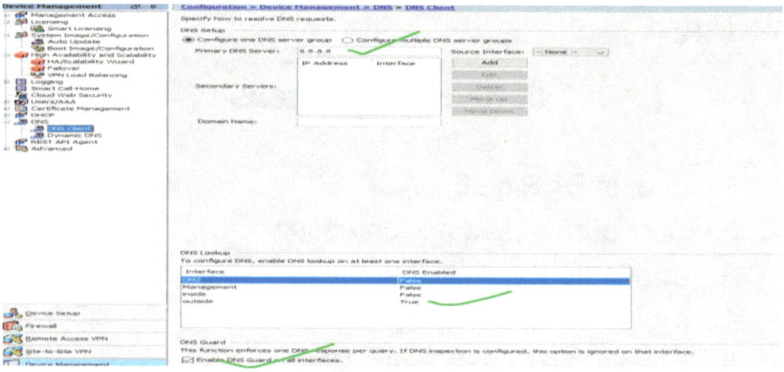

Step 2: Turn on Enable DNS Snooping

Configuration ‣ Firewall ‣ Botnet Traffic Filter ‣DNS Snooping

Under Botnet Traffic Filter, select DNS Snooping.

For global DNS Snooping, simply check the DNS Snooping Enabled option under the global interface.

This procedure enables inspection of DNS packets and enables Botnet Traffic Filter snooping, which compares the domain name with those on the dynamic database or static database, and adds the name and IP address to the Botnet Traffic Filter DNS reverse lookup cache. This cache is then used by the Botnet Traffic Filter when connections are made to the suspicious address.

DNS snooping should only be enabled for DNS traffic. Failure to do so will result in non-DNS traffic being dropped because it is not.

Adhering to the DNS protocol. DNS snooping should only be enabled for the interface that is facing the Internet, since the Botnet Traffic Filter database is aimed at addressing the external threat of botnets.

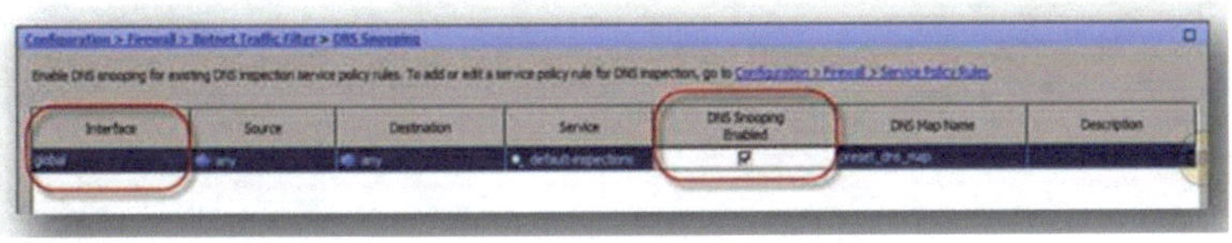

Step-3 Enable the client and use the dynamic database

Configuration ‣ Firewall ‣ Botnet Traffic Filter ‣Botnet Database

Enabling ASA to be a client, this will download all dynamic databases from SIO and then make a decision based on the downloaded dynamic database. Failing this setting, ASA will not have an updated database to verify with.

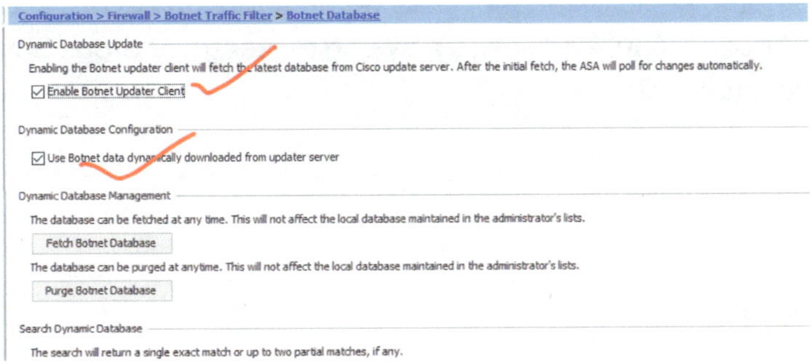

Step-4 Configure the static database (Optional)

Configuration ‣ Firewall ‣ Botnet Traffic Filter ‣blacklist and Whitelist

This procedure lets you augment the dynamic database with domain names or IP addresses that you want to blacklist or whitelist.

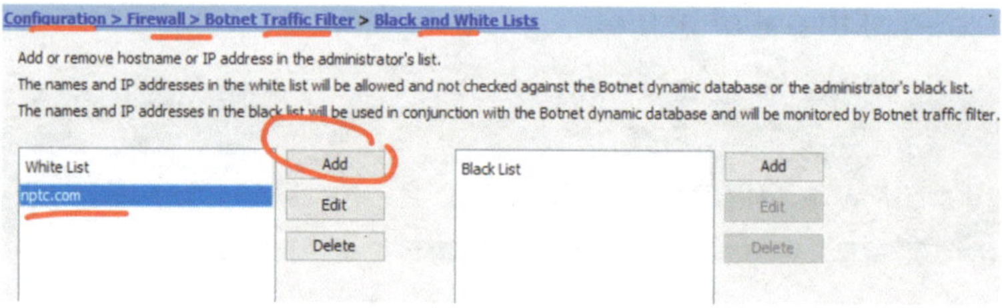

Step-5 Turn on the actions for Botnet Traffic Filter and traffic Classification

Configuration ‣ Firewall ‣ Botnet Traffic Filter ‣Traffic settings

This procedure enables the Botnet Traffic Filter, which compares the source and destination IP address in each initial connection packet to the IP addresses in the dynamic database, static database, DNS reverse lookup cache, and DNS host cache, and sends a syslog message or drops any matching traffic.

Traffic Classification

Given to us dynamically from the database by default is (Medium, High, Very High)

Static black list by default are rated very high.

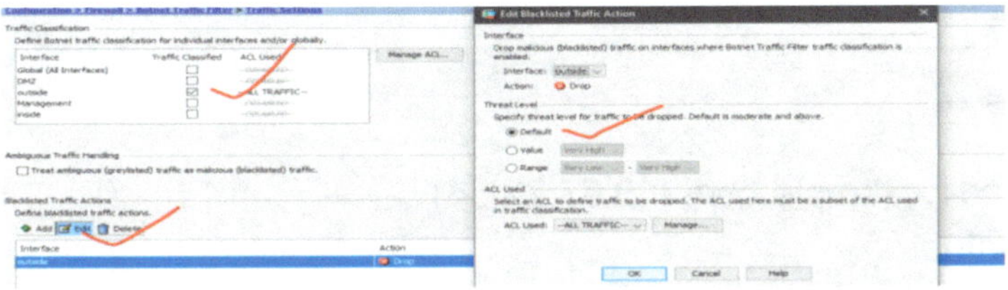

It is also possible to only filter specific traffic; this can be done by selecting Manage ACL and defining the appropriate traffic. It's also possible to specific what level of traffic will be dropped.

Monitoring the Botnet Traffic Filter

Whenever a known address is classified by the Botnet Traffic Filter, a syslog message is generated. You can also monitor Botnet Traffic Filter statistics and other parameters by entering commands on the ASA.

Lab verification result

PrimaryFW/pri/act(config)# show dynamic-filter reports top malware-sites

Malware Sites (since last clear)

Site Connections Logged Dropped Threat-level Category

Sample output from the **show dynamic-filter reports top malware-sites** command in a real-life concept


```
ciscoasa# show dynamic-filter reports top malware-sites

Site                    Connections logged dropped Threat
Level Category.

--------------------------------------------------------------------------------
--------

bad1.example.com (10.67.22.34)          11    0       2
Botnet

bad2.example.com (209.165.200.225)        8    8
3     Virus

bad1.cisco.example (10.131.36.158)        6    6
3     Virus

bad2.cisco.example (209.165.201.1)        2    2
3     Trojan

horrible.example.net (10.232.224.2)      2    2       3
Botnet

nono.example.org (209.165.202.130)        1    1
3     Virus
```

```
ciscoasa# show dynamic-filter reports top malware-ports

Port                            Connections logged
```

----

tcp 1000	617
tcp 2001	472
tcp 23	22
tcp 1001	19
udp 2000	17
udp 2001	17
tcp 8080	9
tcp 80	3
tcp >8192	2

ciscoasa# show dynamic-filter reports top-infected-hosts

Host	Connections logged
10.10.10.51(inside)	1190
10.12.10.10(inside)	10
10.10.11.10(inside)	5

Verification to check on the domain name black listed or whitelisted

PrimaryFW/pri/act(config-llist)# show run dynamic-filter

dynamic-filter updater-client enable

dynamic-filter use-database

dynamic-filter enable interface outside

dynamic-filter drop blacklist interface outside

dynamic-filter whitelist

 name nptc.com

dynamic-filter blacklist

 name bad.com

Impact of Botnet Filtering

Post implementing the entire process of Botnet Filtering, there is a high chance that Botnet Filtering can drop some websites which are most commonly used for business needs. Thus, it is mandatory for us to get an analysis done post the implementation.

- Infrastructure analysis
- Scheduling a downtime
- Run a pilot test
- Recovery action

Recovery Action

Cisco Security Intelligent Operations (CSIO) has a list of updated botnet hackers collected across the globe. It includes websites of various risk levels – low to medium, to high. There is a possibility that some websites which one may need for business are also in the list of the CSIO database, hence get blocked. It is the administrator's responsibility to analyse the risk factors of any given website with their respective infrastructure security team. This will enable one to decide if the website can be added under the static whitelist if it is important for a business.

Virtual Private Network (VPN) - Customer Provision VPNs

A VPN is a technology use to extend access to a private network to users who do not have direct access to it, such as an office network, allowing secure access from off-site over the Internet.

This is achieved by creating a link between computer networks by the use of network tunneling protocols. The goal of a virtual private network is to allow network hosts (PCs, servers, etc.) to exchange network messages across another network to access private content, as if they were part of the same network.

How a VPN Works:

Connection: Your device connects to a VPN server, which is usually located in a different geographical location.

Encryption: The VPN encrypts all your internet traffic before it leaves your device.

Tunnelling: The encrypted data is sent through a secure tunnel to the VPN server.

Decryption: The VPN server decrypts the data and sends it to its intended destination.

Masked IP Address: Your IP address is masked by the VPN server, so your original IP address is not visible.

Benefits of Using a VPN:

Enhanced Privacy:

VPNs encrypt your data and hide your IP address, making it harder for third parties to track your online activities and personal information.

Increased Security:

VPNs protect your data when using public Wi-Fi networks, which are often insecure.

Access to Geographically Restricted Content:

VPNs can bypass regional restrictions and allow you to access content from different locations.

Secure Remote Access:

VPNs allow users to access private networks and company resources securely from remote locations.

Types of VPN.

Remote Access: This method uses a host to network; this type of extension provides computer access to the local area network of a remote site, or any wider enterprise networks.

This may be employed for remote workers or to enable people to access their private company resources without exposing them to the public Internet.

Remote-access VPNs, which are typically user-initiated, may use passwords, biometrics, two-factor authentication, or other cryptographic methods.

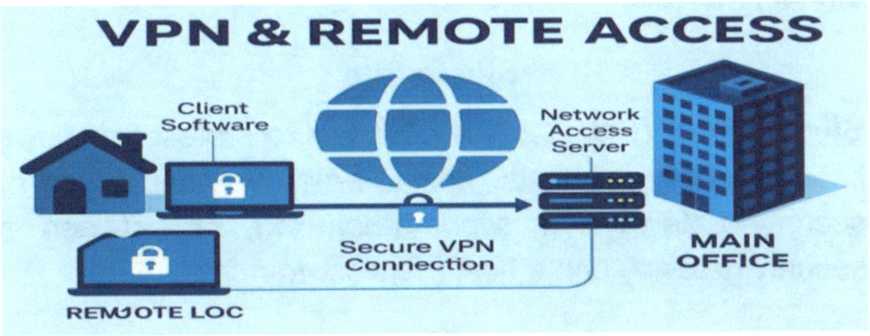

Site-to-site: This method is use to connect two networks. Tunneling is only done between two devices (like routers, firewalls, VPN Concentrators, Servers, etc.) located at both network locations.

Businesses tend to make use of site-to-site connections for business-to-business, cloud computing, and branch office scenarios.

Site-to-site VPNs often use passwords (pre-shared keys) or digital certificates. Depending on the VPN protocol, they may store the key to allow the VPN tunnel to establish automatically, without intervention from the administrator.

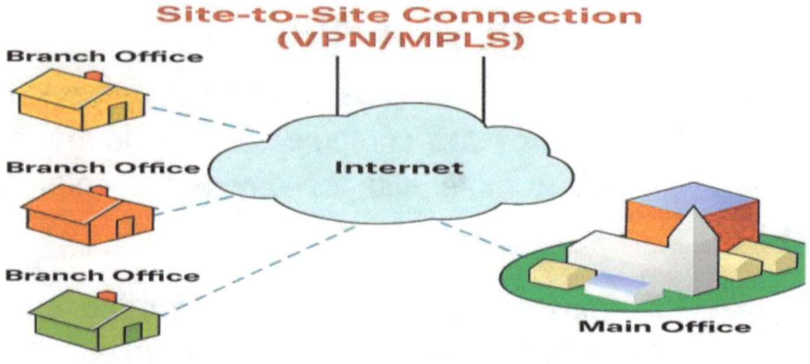

Technologies Used By VPN

A virtual private network is based on a tunneling protocol. Cisco ASA support two types of Internet Protocol Security (IPsec) VPN and Secure Socket Tunnel SSL/TLS VPN tunnel protocols.

Site-to-Site

Site-to-Site VPN uses IPsec tunnels **to** protect data flows between a pair of hosts (*host-to-host*), between a pair of security gateways (*network-to-network*), or between a security gateway and a host (*network-to-host*).

Remote Access VPN

Remote access can use either SSL or IPsec technology for secure remote connections.

SSL VPN- There are mainly two types of SSL VPNs supported by Cisco devices

Clientless Mode WebVPN: This is the first implementation of SSL WebVPN supported. Let's users establish a secure remote access VPN tunnel using just a Web browser. There is no need for a software or hardware VPN client. However, only limited applications can be accessed remotely.

Any Connect WebVPN: A special Java-based client is installed on the user's computer. Providing an SSL secure tunnel to the central site. Provides full network connectivity (similar to an IPsec Remote Access client). All applications and network resources at the central site can be accessed remotely.

Implementing SSL AnyConnect VPN on Cisco ASA

Cisco AnyConnect is a Virtual Private Network (VPN) client software that allows users to establish a secure, encrypted connection to a corporate network from off-campus locations.

Key Features and Functionality:

Secure VPN Connection:

AnyConnect uses encryption protocols like SSL and IPSec to create a secure, private tunnel for data transmission between the user's device and the corporate network.

Client-Side Software:

It's a software application that users install on their devices (Windows, macOS, iOS, Android).

Authentication Options:

It supports various authentication methods, including username/password, two-factor authentication, and digital certificates.

Network Roaming:

AnyConnect can seamlessly re-establish the VPN connection after network changes or device standby.

Use Cases Scenarios:

Remote Access to Corporate Resources:

Allows users to access email, file servers, databases, and other restricted resources from off-campus locations.

Secure Remote Work:

Enables employees to work securely from anywhere by connecting to the corporate network.

University/School Access:

Provides students and faculty with secure access to university networks and resources.

Secure Application Connectivity:

Allows users to access specific applications on the corporate network, even if the applications are not directly exposed to the internet.

SSL AnyConnect VPN Project Task

Project Task: Create an SSL AnyConnect VPN client using the Topology below.

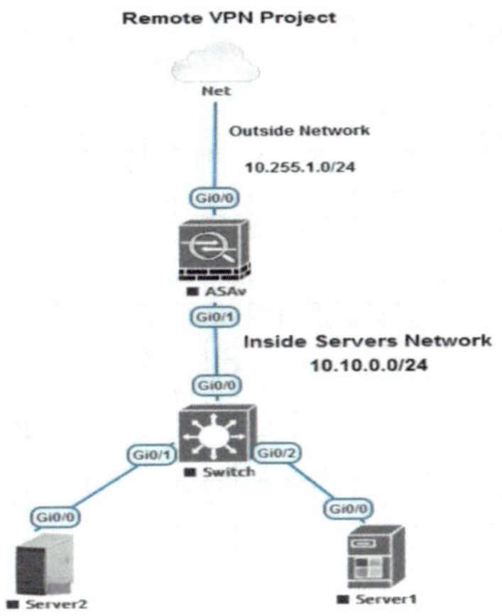

Step 1. Use the wizard to create a new SSL AnyConnect

Define the connection profile name

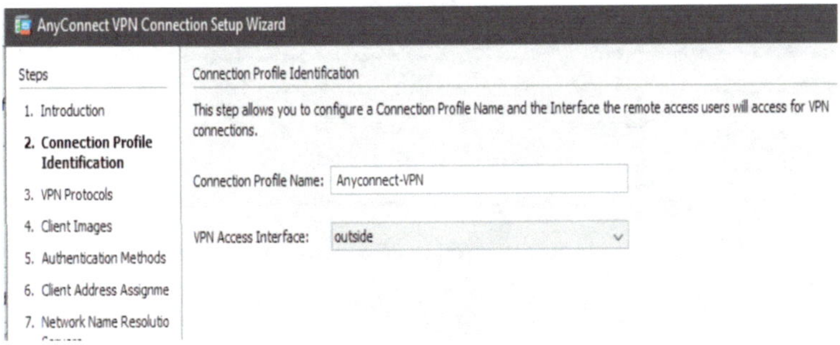

Step 2: Select which protocol you want to use to protect the data, either IPsec or SSL and select the device certificate if any.

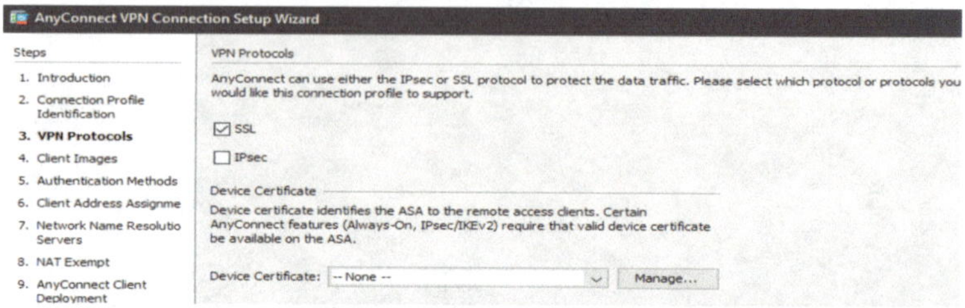

Step 3: Allow ASA to automatically upload the latest AnyConnect package to the client.

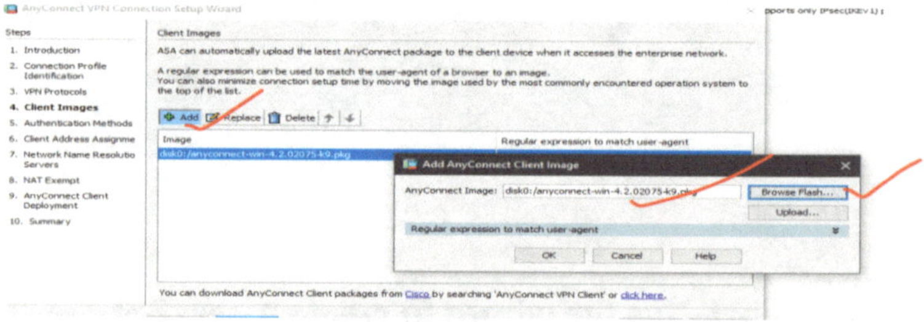

Step 4: Create the authentication method either with a AAA server or use LOCAL and create an account if it is local.

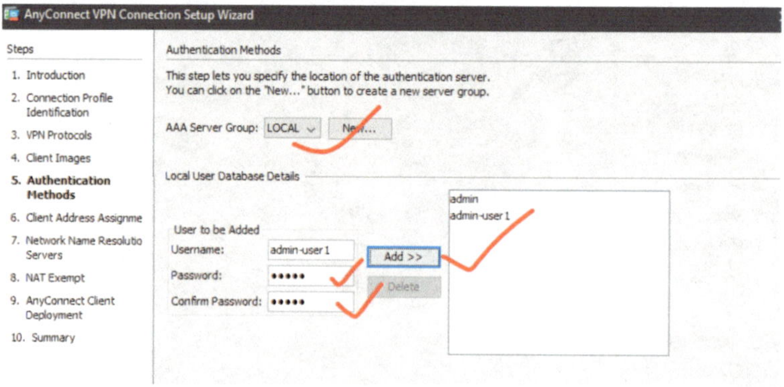

Step 5: Create a DHCP pool for the remote vpn client.

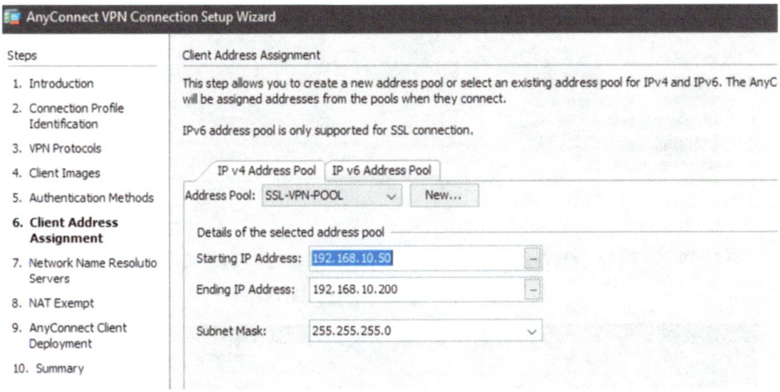

Step 6: Add your DNS server info (should be corporate info)

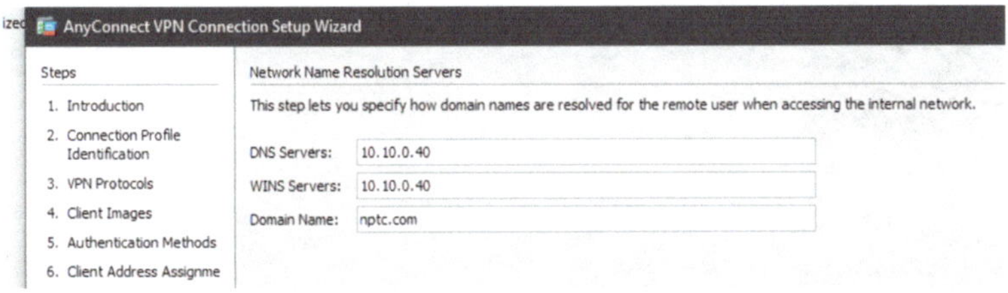

Step 7: Exempt NAT from vpn traffic if NAT is enabled on this ASA

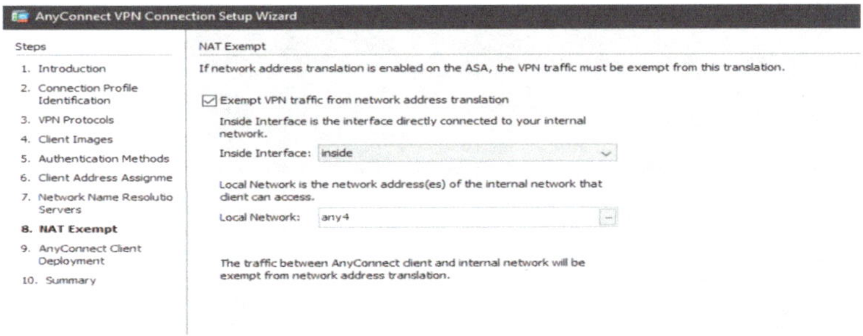

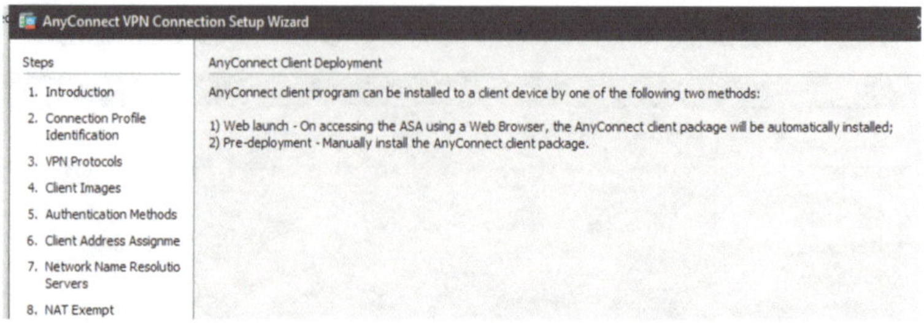

Step 8: Go through the summary of the configuration before clicking finish

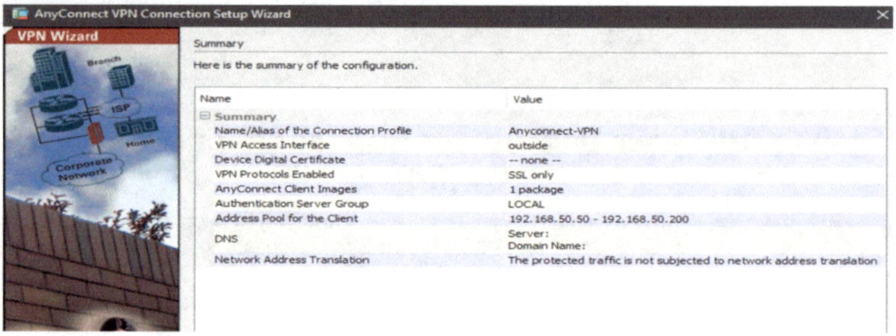

Step 9: Ensure all these are checked with your connection profile

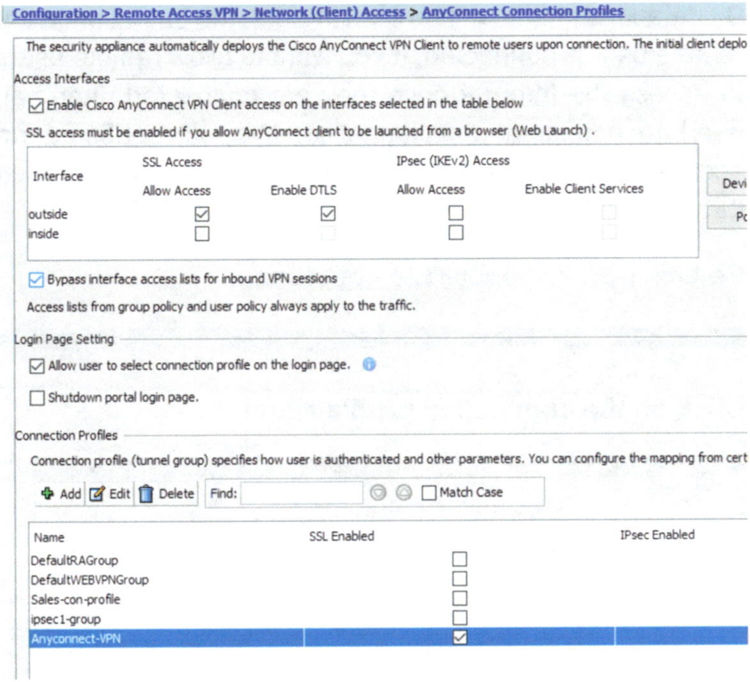

Split Tunnel

By default, all traffic will be sent through the tunnel once the remote user is connected. If you want to allow remote users to access the Internet once they are connected, then you need to configure split tunneling. We will configure an access list that specifies what networks we want to reach through the tunnel.

So let's now configure the split tunnel

Click on the connection profile > edit

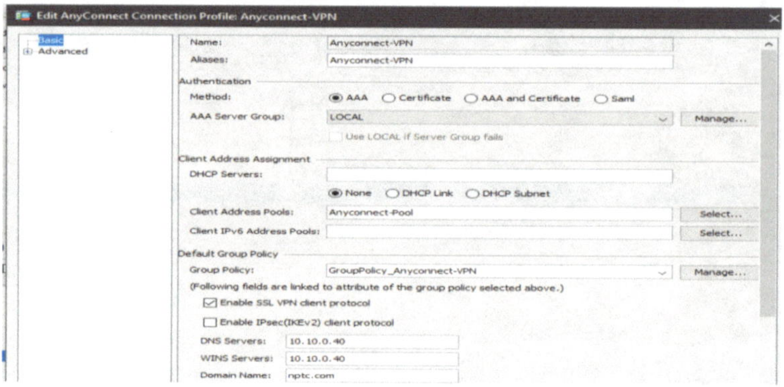

Click on manage under default.

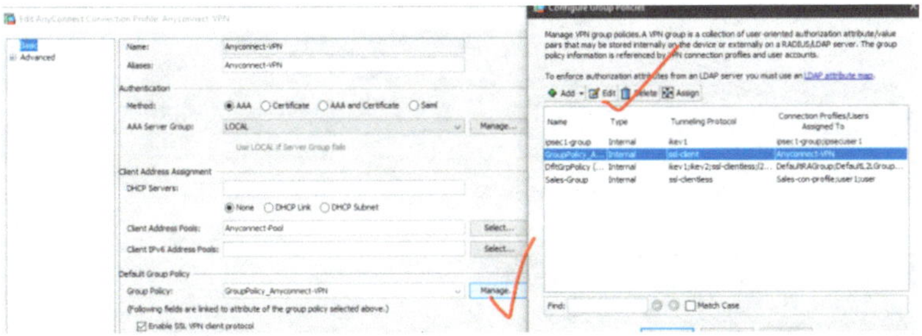

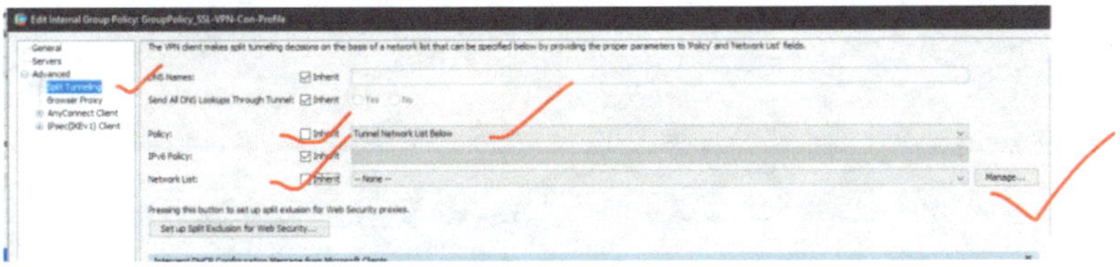

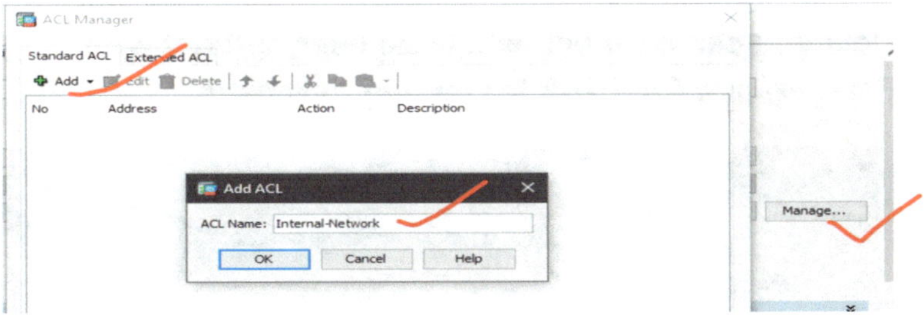

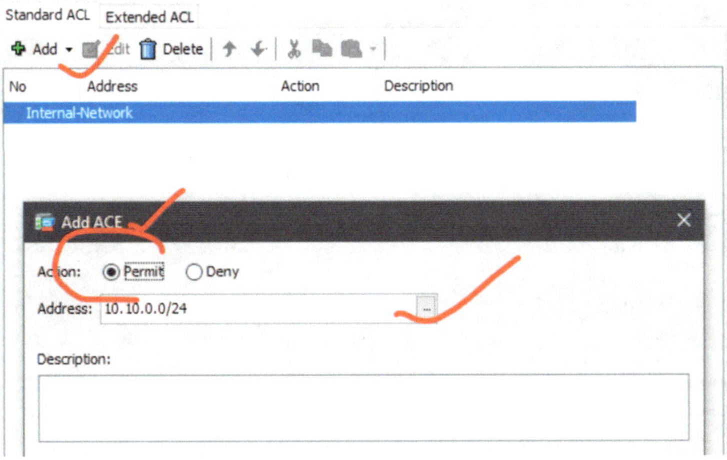

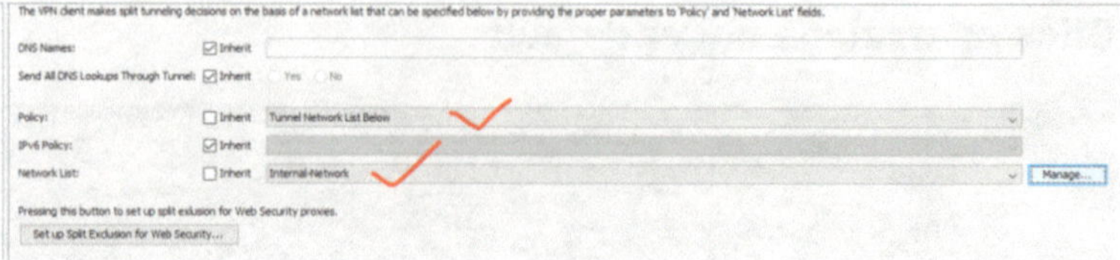

Now we can test our vpn.

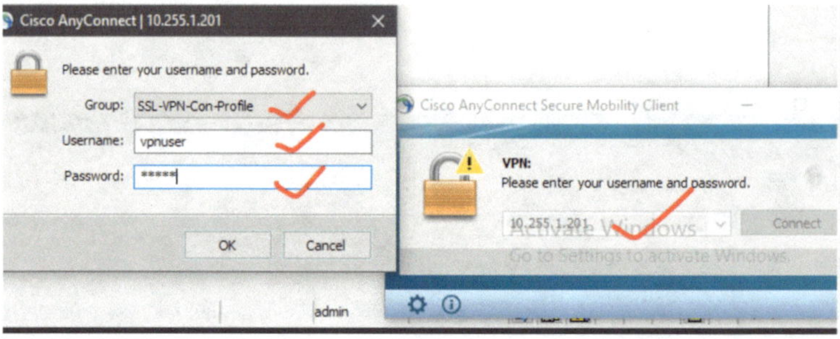

Now you can connect as a local user, but not with RDP, so use AnyDesk for testing purposes.

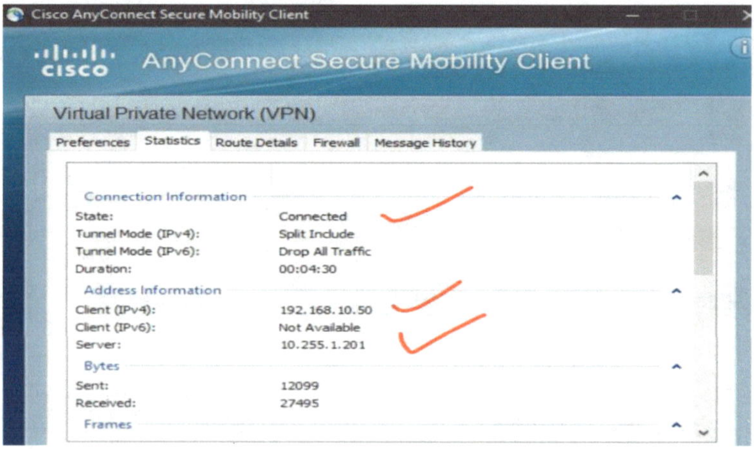

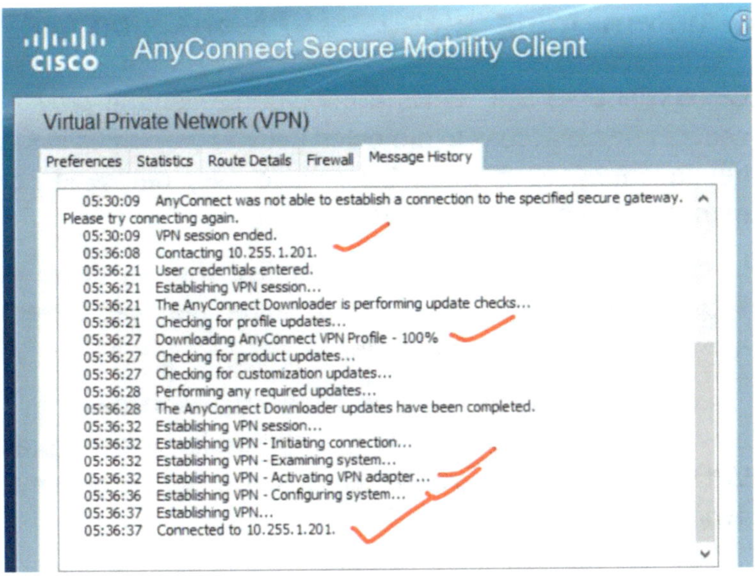


```
ciscoasa# show vpn-sessiondb summary
-----------------------------------------------------------------------
----
VPN Session Summary

-----------------------------------------------------------------------
----
                              Active : Cumulative : Peak Concur : Inac
tive
                             ------------------------------------------
----
AnyConnect Client            :      1 :          4 :            1 :
   0
  SSL/TLS/DTLS               :      1 :          4 :            1 :
   0
```

Implementing Clientless VPN on Cisco ASA

Cisco clientless SSL VPNs allow organizations to provide secure remote access to protected network resources in the headquarters, even when the remote user device is not managed or has no VPN client installed.

In other words, it provides the simplest way for users to access mainly web-based (and some non-web-based) applications over a web browser.

The VPN gateway that acts as a proxy between the remote user and protected resources is responsible for the overall VPN permissions, such as services allowed, bookmarks available

Keep in mind that the Cisco Firepower NGFW firewall doesn't support clientless VPN deployment.

Typical use cases include internet kiosks, on-demand, and business partners that require access only to a specific set of services and resources in general, which works perfectly with the clientless VPN limitation.

However, the clientless TLS VPN solution has some limitations. Because everything is done through the web portal, it may require user training so that users can learn how to use the navigation portal before they begin using it.

Furthermore, due to its proxying nature, real-time applications often experience latency and delay, which makes them unusable at times.

Finally, since this VPN solution doesn't support all IP applications, sometimes you have to choose a different option.

Clientless SSL VPN on ASA Project

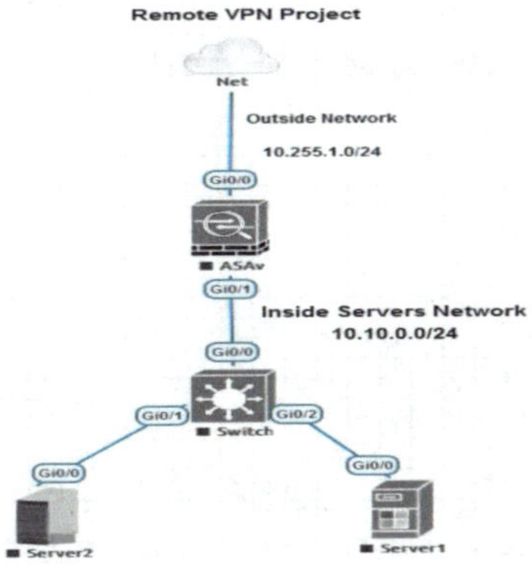

Task: The project requirement is to create a clientless VPN where any random machine on the internet can connect to the internal sales server for corporate work

The **clientless WebVPN** method does not require a VPN client to be installed on the user's computer. You just open your web browser, enter the IP address of the ASA, and you will get access through a web portal. You only have limited access to a number of applications, for example:

- Internal websites (HTTP and HTTPS)
- Web applications
- Windows file shares
- Email servers (POP3, IMAP, SMTP)
- Microsoft Outlook Web Access

There is **no full network access** when you use clientless WebVPN

Step 1: Create a new vpn policy for sales users

Configuration▸Remote Access ▸ Clientless SSL VPN Access ▸ Group Policies ▸ Add

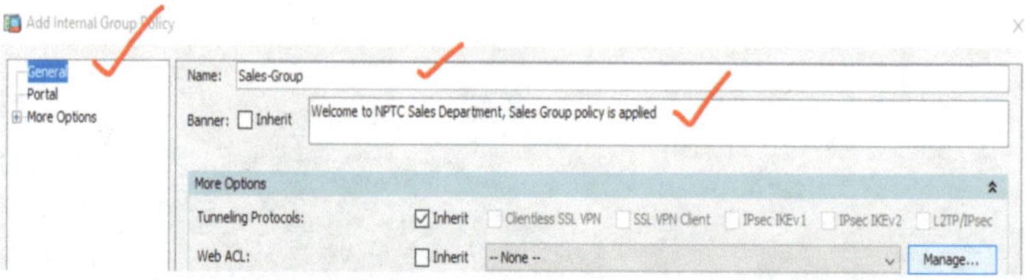

Step 2: Apply web type ACL to prevent sales users from accessing certain services (Optional)

1. First, create a web ACL
2. Second, Create ACE – Access Control Entry

First, create a web ACL

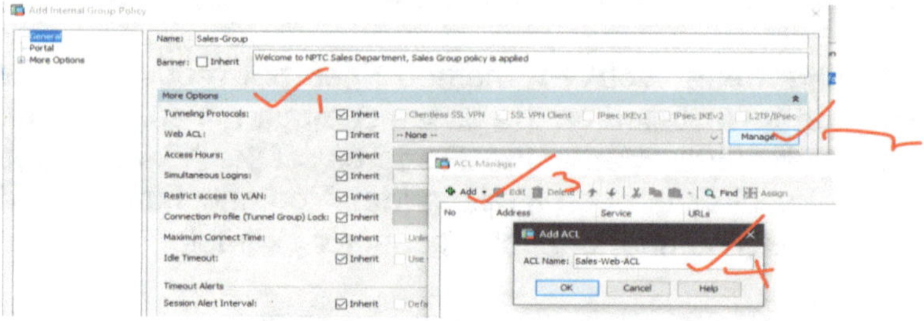

Second, Create ACE – Access Control Entry

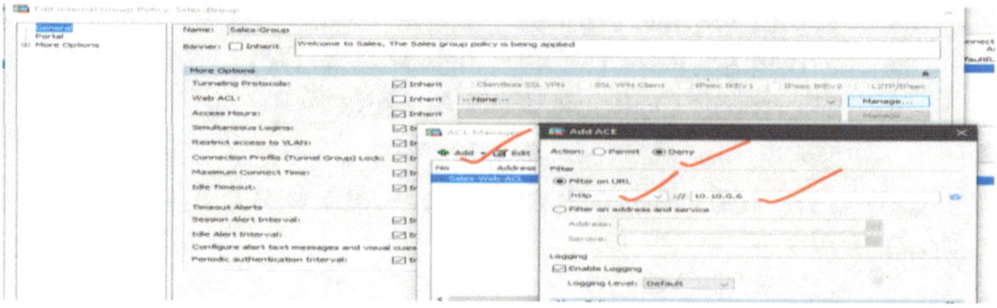

Now permit all other traffic without this, and everything will be dropped.

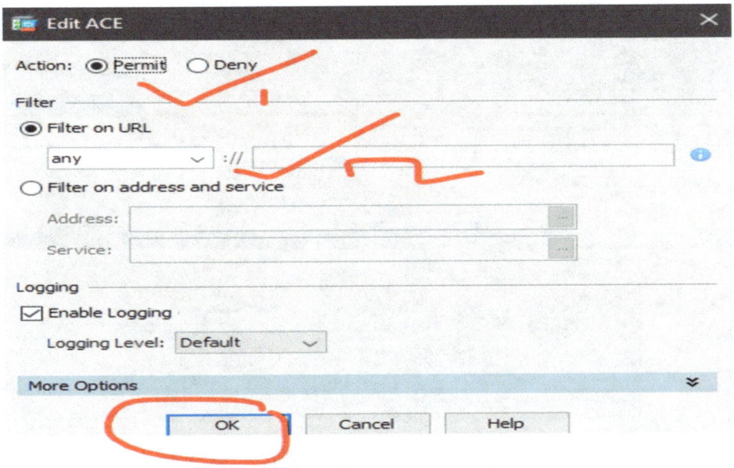

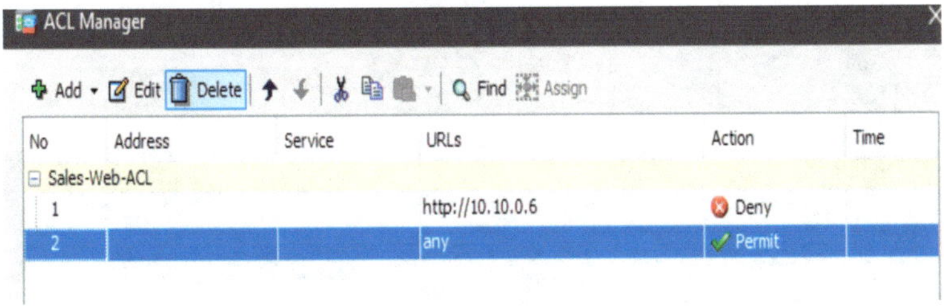

The last step for the ACL is to select it from the drop-down menu.

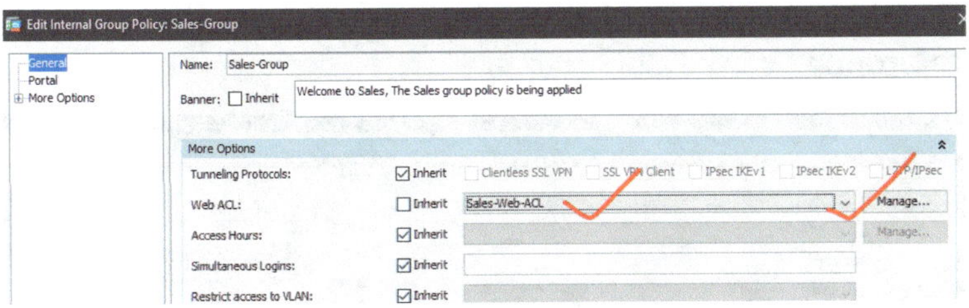

Step 3: Create the bookmark list.

Add internal Group Policy▸ Portal ▸ un-chick Bookmark list▸ Manage ▸ Add

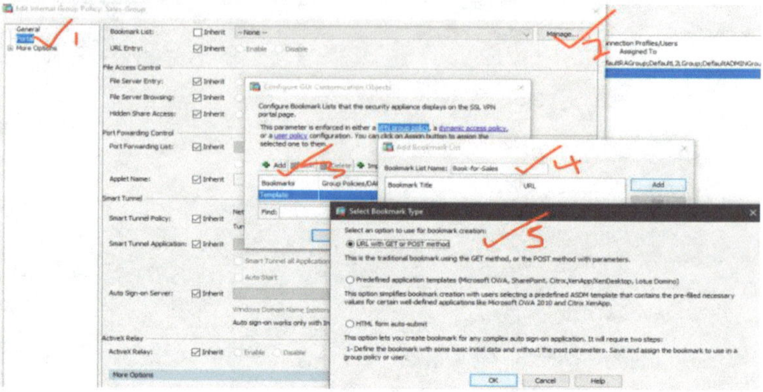

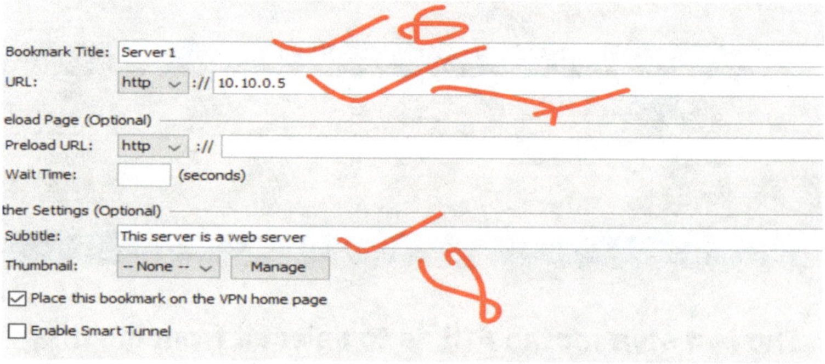

Creating another Bookmark starting from item 6 as a reference.

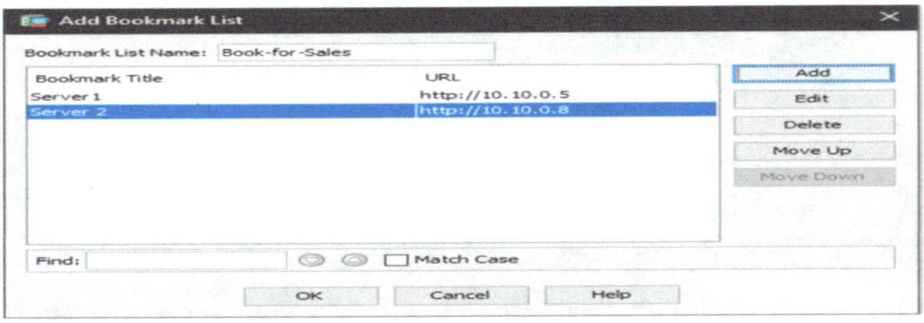

Enable the URL Entry on the ASA under the portal.

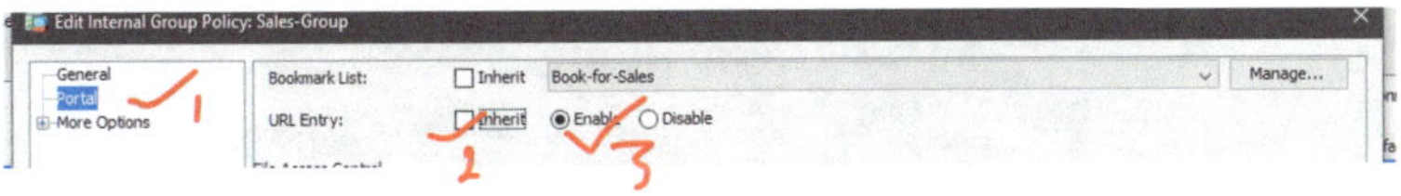

The Final Step is to click Apply for all the group policy configurations to take effect.

Step 4: Create the Connection profiles to be used for this group.

Configuration▸Remote Access ▸ Clientless SSL VPN ▸ Connection Profile ▶ Add

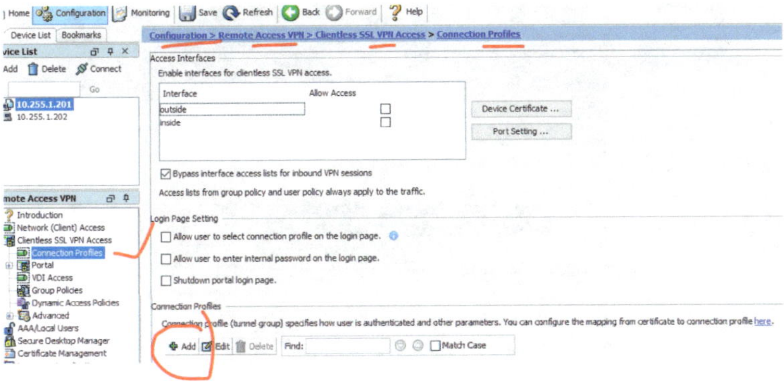

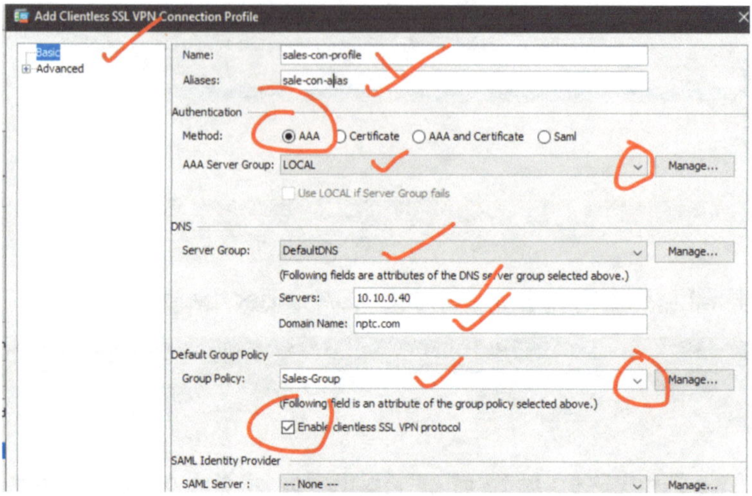

Now, create the URL that the users are going to connect with.

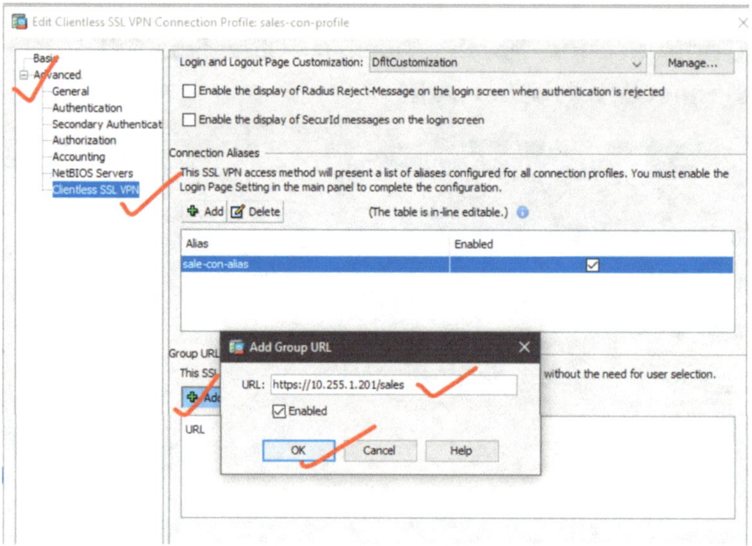

Enable clientless SSL VPN traffic termination on Cisco ASA's interface where the remote sessions will arrive.

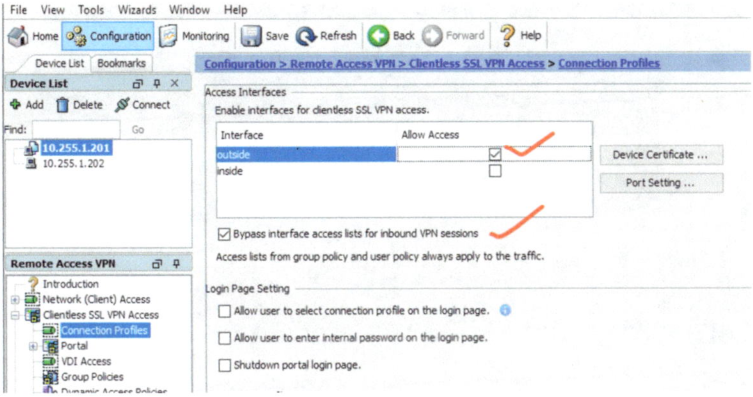

The final stage is to click Apply for the configuration to take effect.

Let's create users who can access this VPN by using cisco as a password.

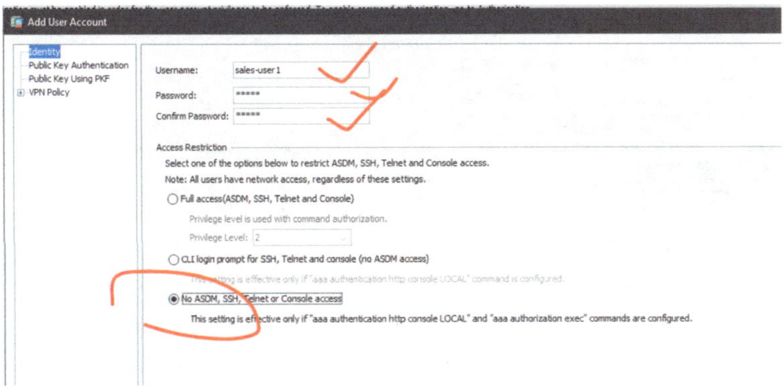

Add the User to the right vpn policy group and connection profile group.

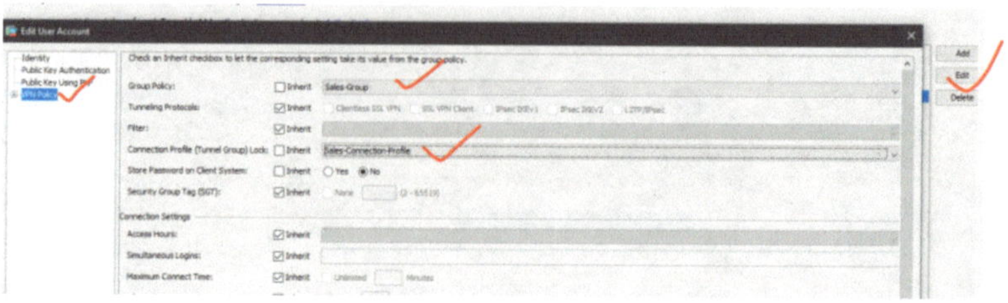

Let's verify

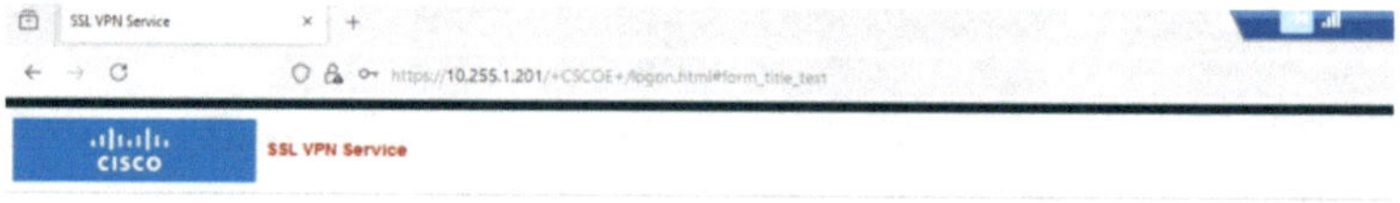

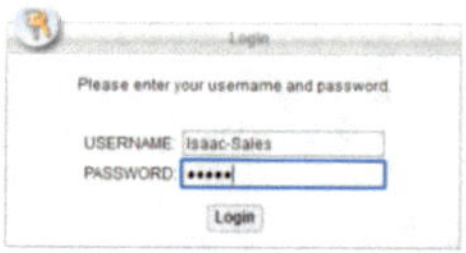

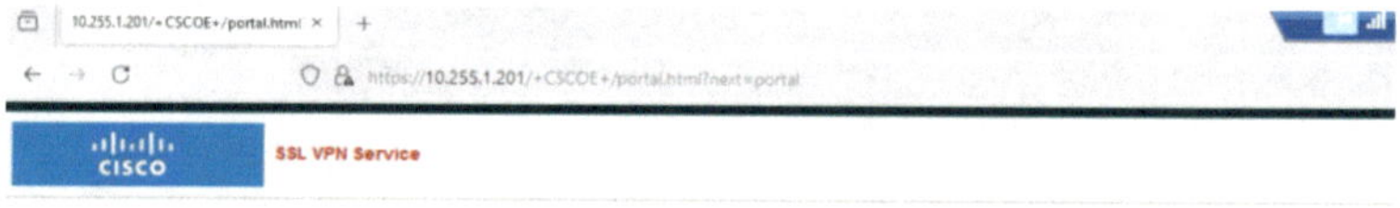

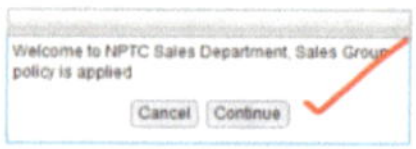

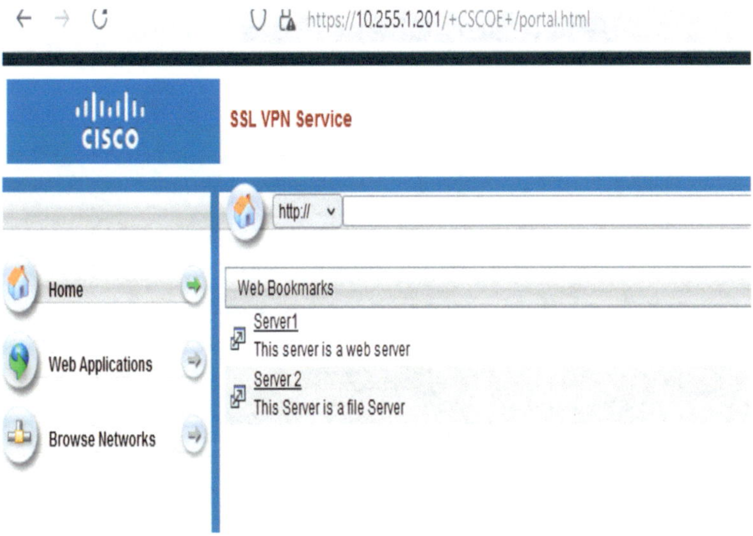
https://10.255.1.201/+CSCOE+/portal.html
SSL VPN Service
http://
Home
Web Applications
Browse Networks
Web Bookmarks
Server1
This server is a web server
Server 2
This Server is a file Server

VPN IPSEC Client Project Task

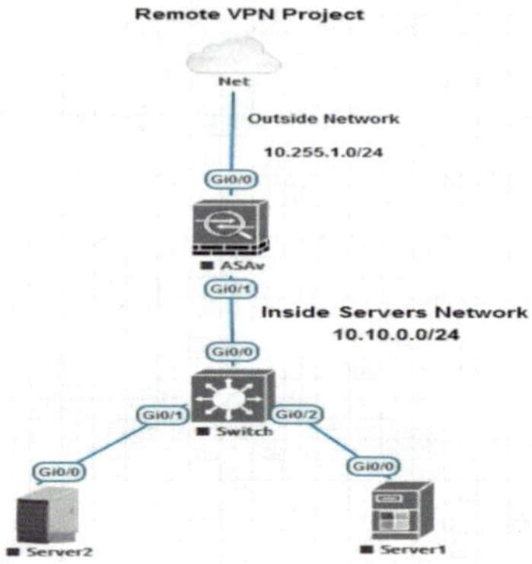

Task: Create an IPsec VPN client using the Topology above

Step 1: Use the wizard to create a new IPsec (IKEV1)

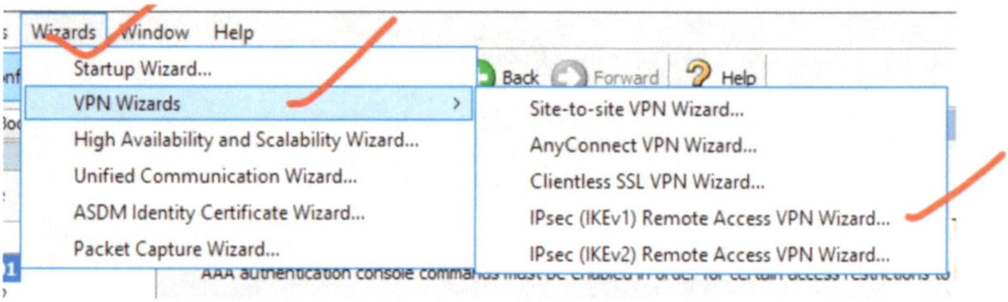

Step 2. Choose the interface that users can connect with

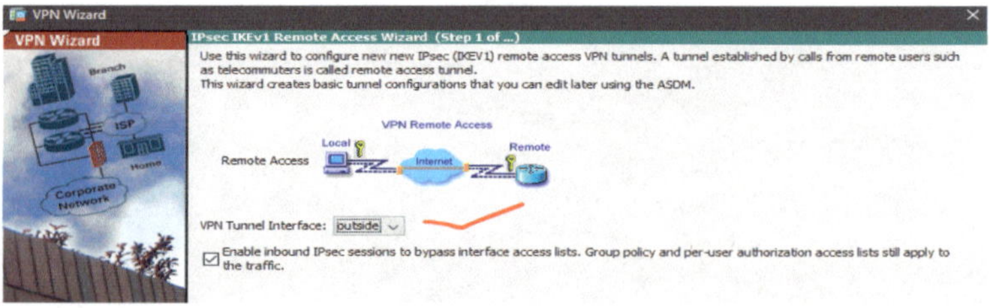

Step 3: Select the type of VPN client type

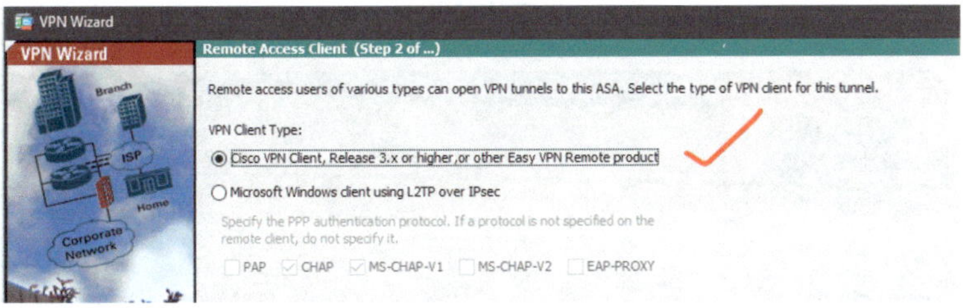

Step 4: Choose the authentication method and the tunnel group name.

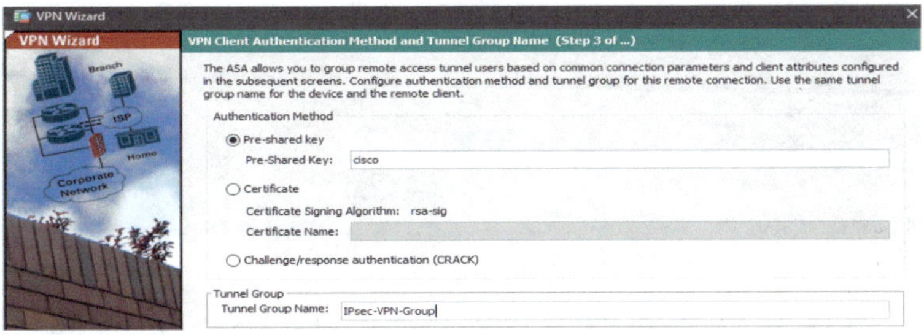

Step 5: Choose Client Authentication Method, either with AAA server or local user.

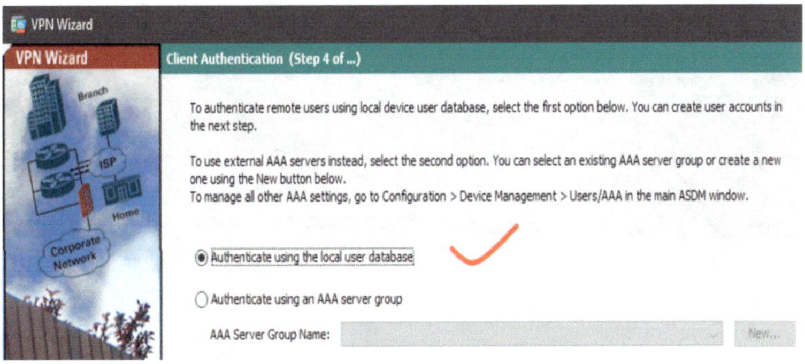

Step 6: Create the DHCP pool to be used by the VPN client.

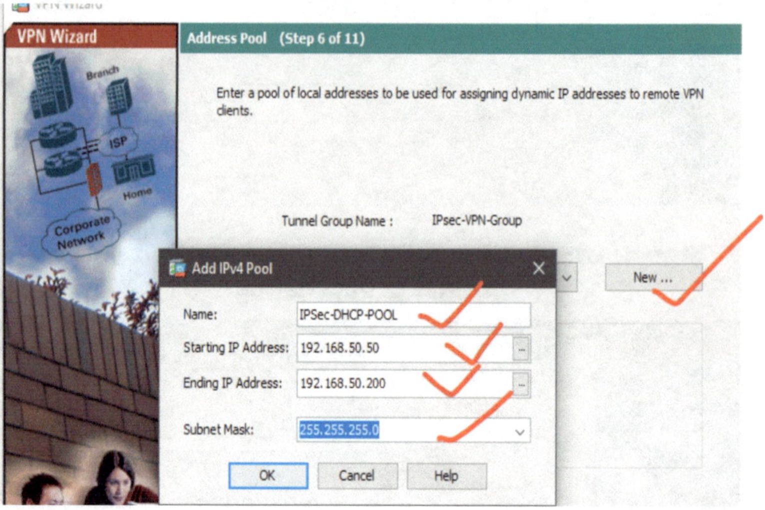

Step 7: Attribute the client VPN with DNS services.

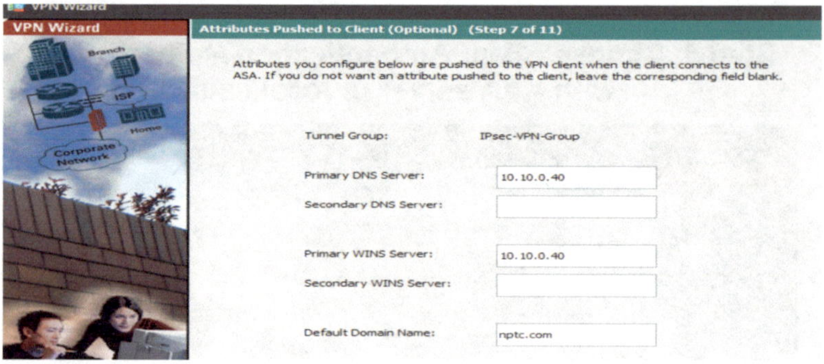

Step 8: Create exceptions to NAT for remote users and enable split tunneling for remote users as well.

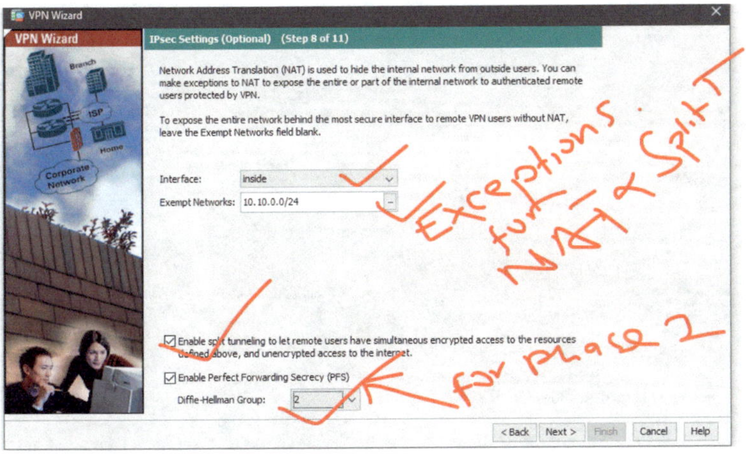

Step 9: Go through the Summary of what you created so far before you click finish.

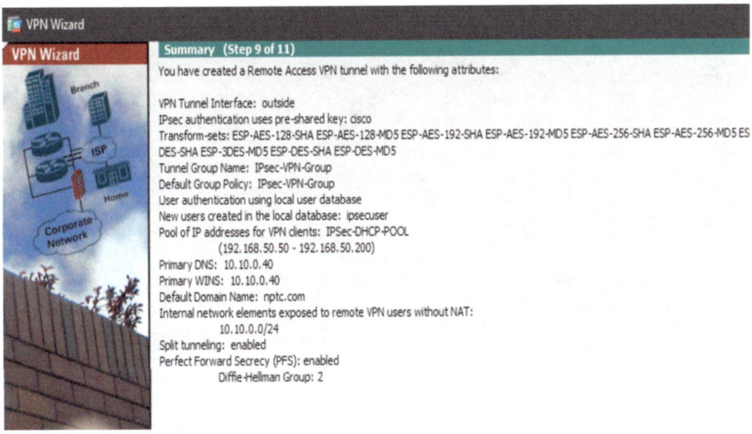

Step 10: Ensure that under the connection profile, IPsec is enabled.

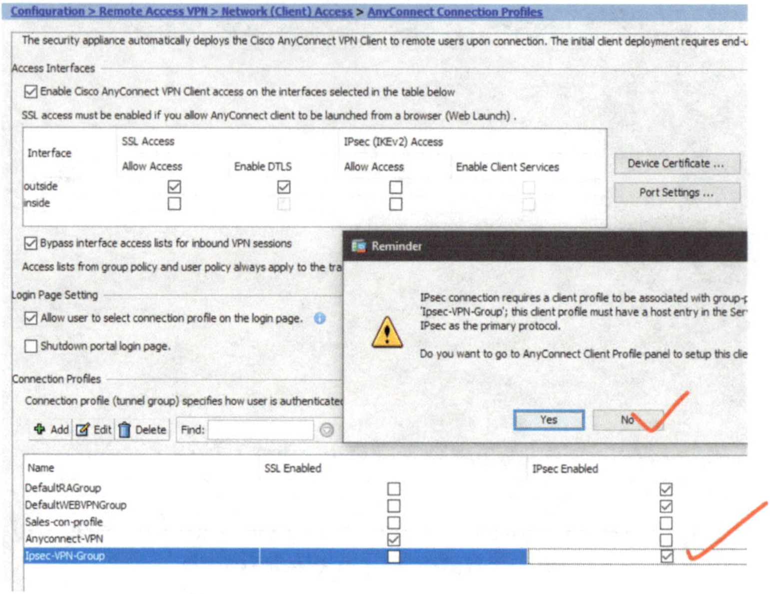

Step 11: Configure the client vpn on the client device. IPsec is not supported on Windows 10, so we use a third-party.

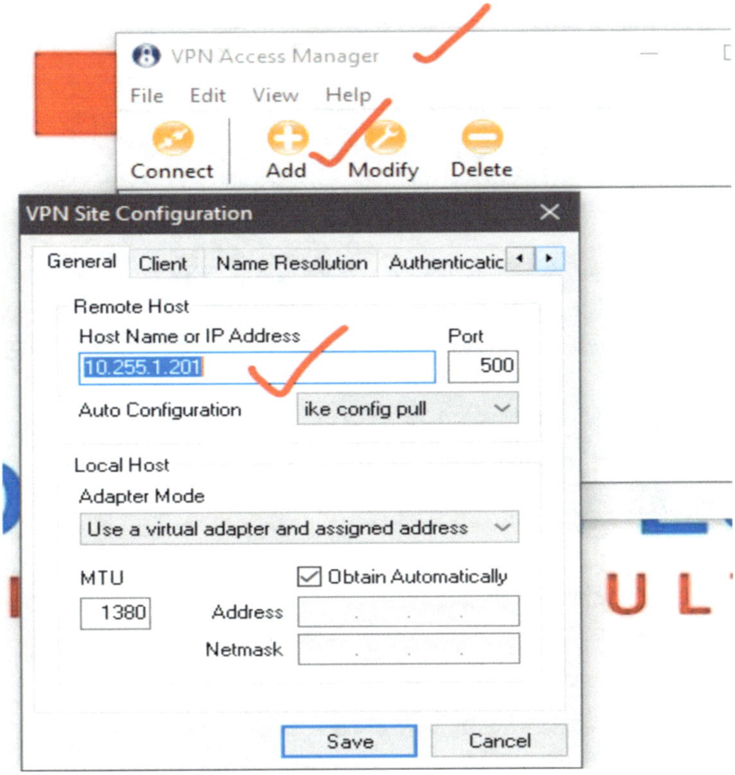

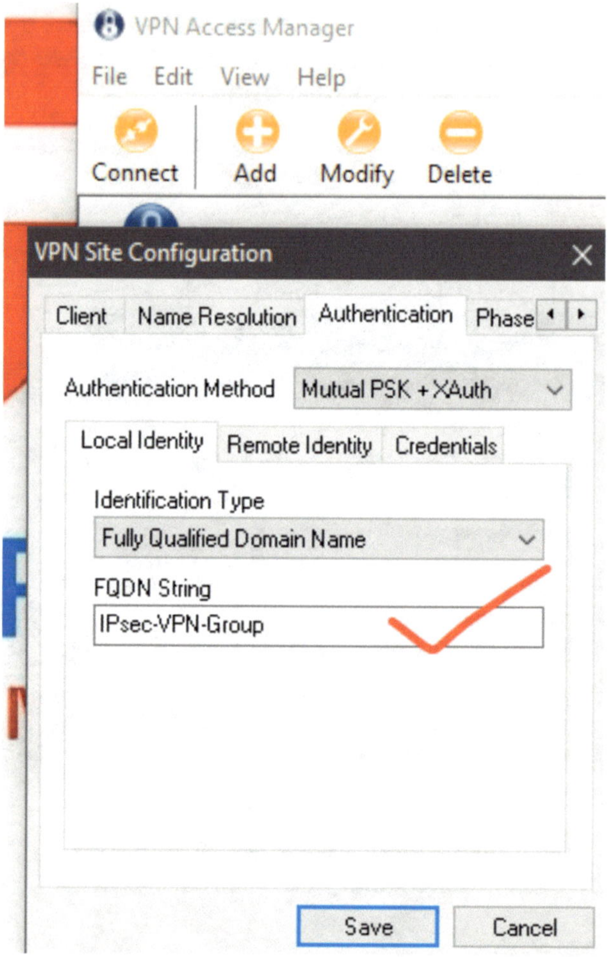

VPN Access Manager
File Edit View Help
Connect
Add
Modify
Delete
VPN Site Configuration
Client
Name Resolution
Authentication
Phase
Authentication Method
Mutual PSK + XAuth
Local Identity
Remote Identity
Credentials
Identification Type
Fully Qualified Domain Name
FQDN String
IPsec-VPN-Group
Save
Cancel

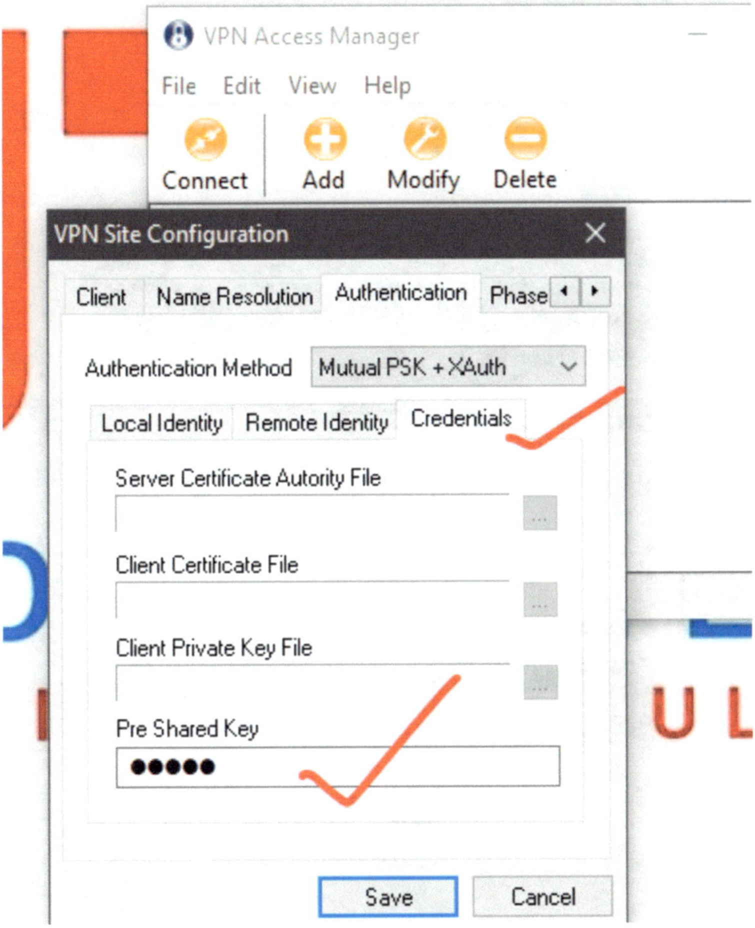
VPN Access Manager
File Edit View Help
Connect
Add
Modify
Delete
VPN Site Configuration
Client
Name Resolution
Authentication
Phase
Authentication Method
Mutual PSK + XAuth
Local Identity
Remote Identity
Credentials
Server Certificate Autority File
Client Certificate File
Client Private Key File
Pre Shared Key
Save
Cancel

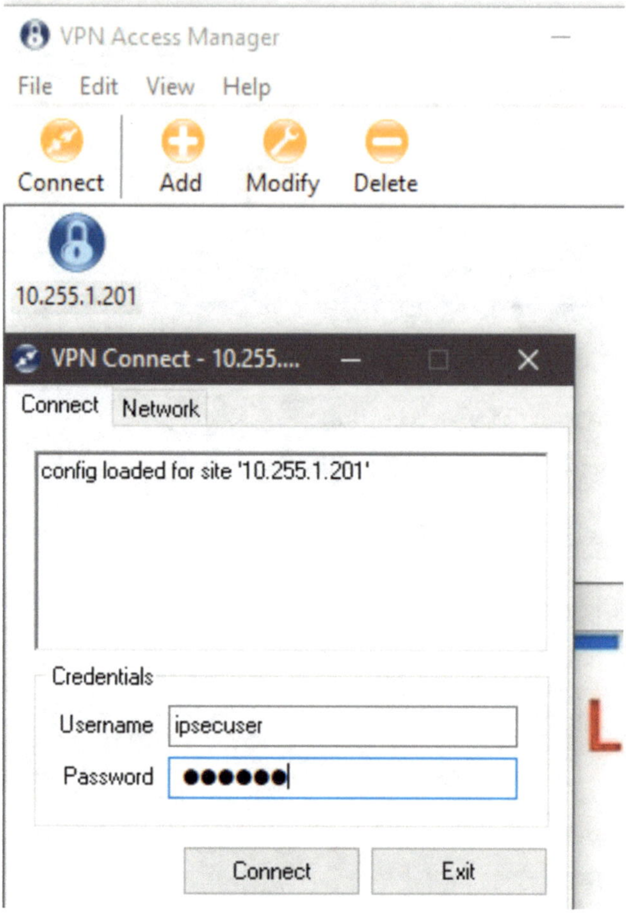
VPN Access Manager
File Edit View Help
Connect
Add
Modify
Delete
10.255.1.201
VPN Connect - 10.255....
Connect
Network
config loaded for site '10.255.1.201'
Credentials
Username
ipsecuser
Password
Connect
Exit

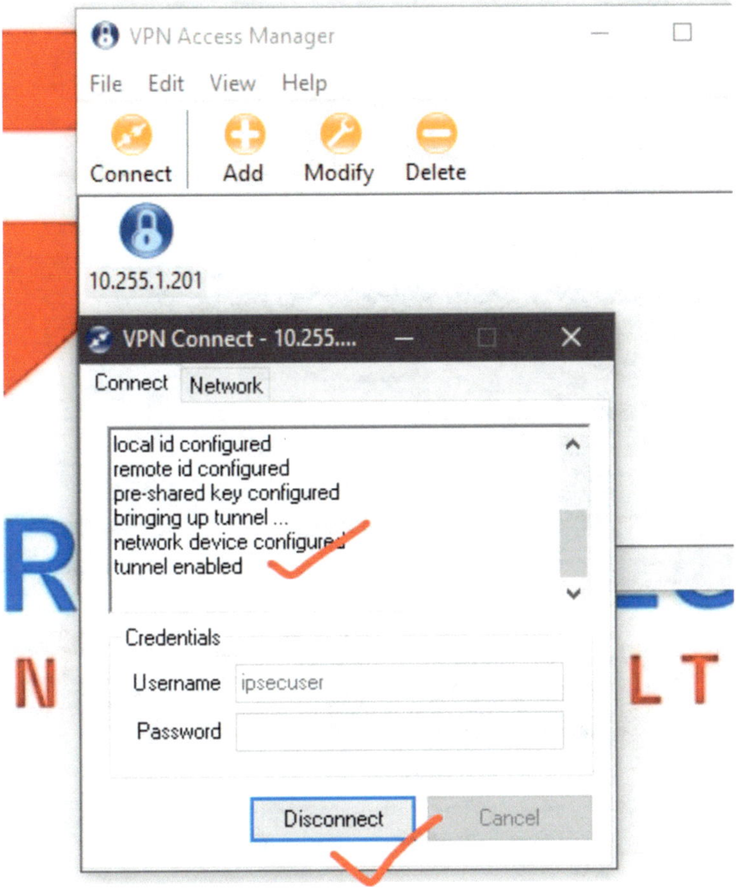
VPN Access Manager
File Edit View Help
Connect
Add
Modify
Delete
10.255.1.201
VPN Connect - 10.255....
Connect
Network
local id configured
remote id configured
pre-shared key configured
bringing up tunnel ...
network device configured
tunnel enabled
Credentials
Username
ipsecuser
Password
Disconnect
Cancel

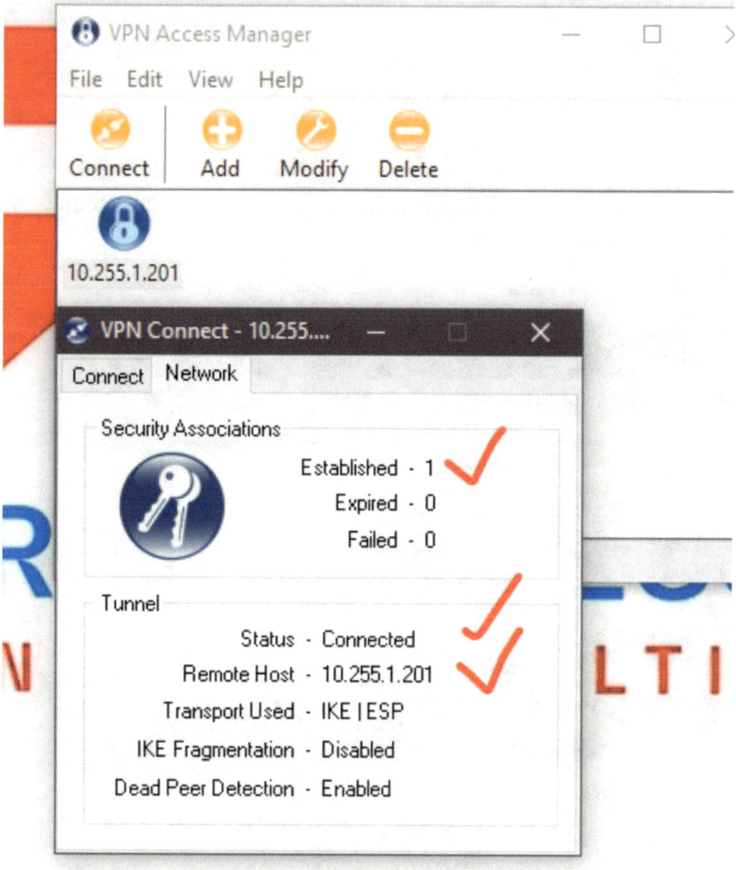


```
ciscoasa# show vpn-sessiondb summary
---------------------------------------------------------------------------
VPN Session Summary
---------------------------------------------------------------------------
                              Active : Cumulative : Peak Concur : Inactive
                            ----------------------------------------------
AnyConnect Client           :      0 :          1 :           1 :        0
  SSL/TLS/DTLS              :      0 :          1 :           1 :        0
IKEv1 IPsec/L2TP IPsec      :      1 :          2 :           1
---------------------------------------------------------------------------
Total Active and Inactive   :      1             Total Cumulative :      3
Device Total VPN Capacity   :    250
Device Load                 :     0%
---------------------------------------------------------------------------
```

Implementing Site-to-Site IPSEC VPN

Site-to-Site IPSec VPN Tunnels are used to allow the secure transmission of data, voice, and video between two sites (e.g, offices or branches). The VPN tunnel is created over the public Internet and encrypted using a number of advanced encryption algorithms to provide confidentiality of the data transmitted between the two sites.

IPsec VPN Requirements

Phase 1 Configuration and Its Purpose

Phase 1 of IPsec is used to establish a secure channel between the two peers that will be used for further data transmission. The ASAs will exchange secret keys, they will authenticate each other, and they will negotiate about the IKE security policies. This is what happens in phase 1:

- Authenticate and protect the identities of the IPsec peers.
- Negotiate a matching IKE policy between IPsec peers to protect the IKE exchange.
- Perform an authenticated Diffie-Hellman exchange to have matching shared secret keys.
- Set up a secure tunnel for IKE phase 2.

Phase 2 Configuration and Its Purpose

Once the secure tunnel from phase 1 has been established, we will start phase 2. In this phase, the two firewalls will negotiate the IPsec security parameters that will be used to protect the traffic within the tunnel. In short, this is what happens in phase 2:

- Negotiate IPsec security parameters through the secure tunnel from phase 1.
- Establish IPsec security associations.
- Periodically renegotiates IPsec security associations for security.

Site-to-Site IPsec VPN Project

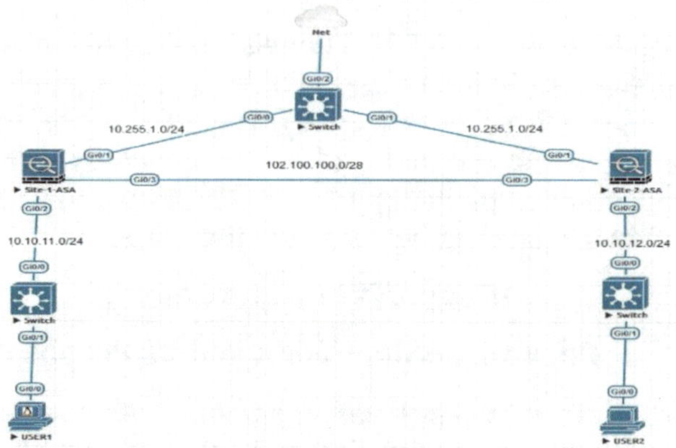

Task; Create an IPsec site-to-site VPN using the Topology above

Step 1: Open your ASDM for Site-1 > click Wizards > VPN wizard>site to site vpn> Next

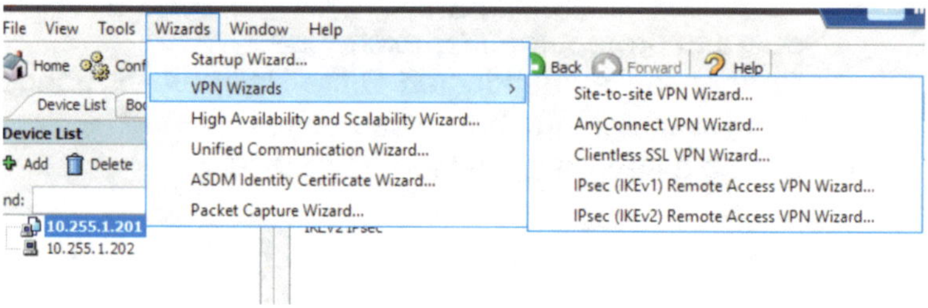

Step 2: Enter your Peer IP address

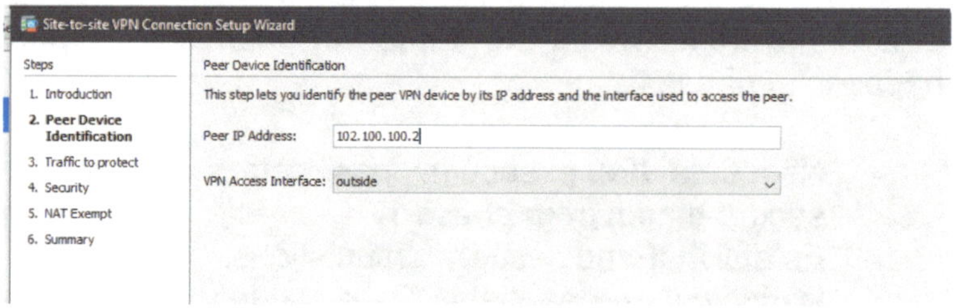

Step 3: Select which traffic to encrypt

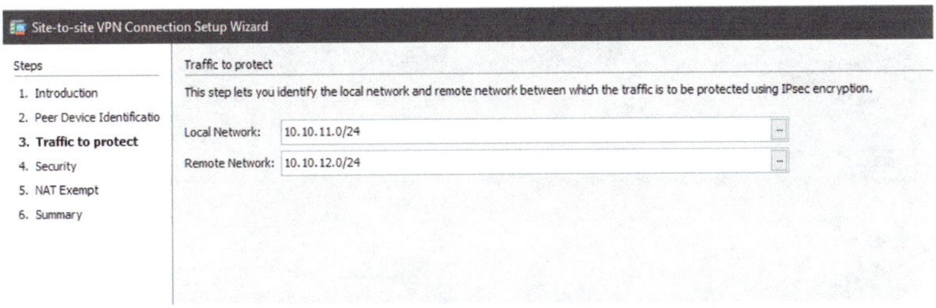

Step 4: Specify the PSK as cisco

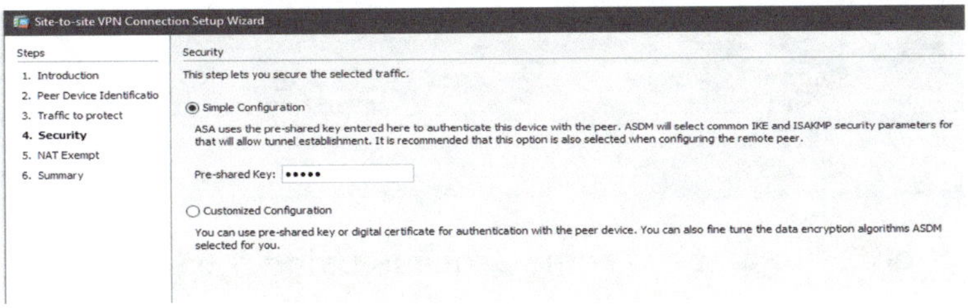

Step 5: Exempt NAT rule for the local address

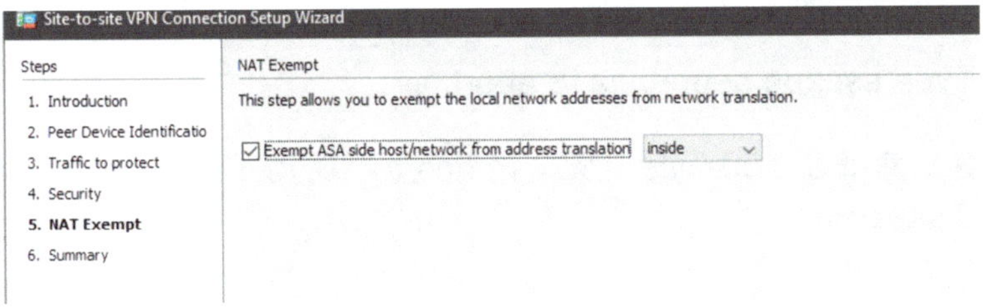

Step 6: Go through the summary and edit when needed

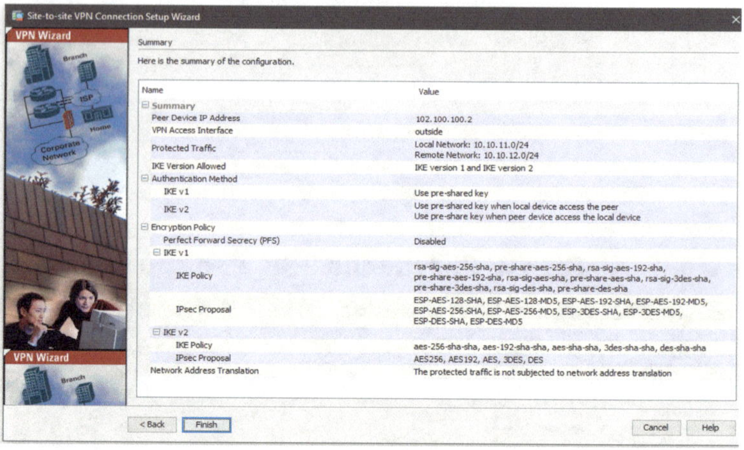

Step 7: Repeat the same process using the same PSK and your Peer IP address

Step 8: Let's test the tunnel by pinging from a user

USER1#ping 10.10.12.2

Type escape sequence to abort.

Sending 5, 100-byte ICMP Echos to 10.10.12.2, timeout is 2 seconds:

.!!!!

Success rate is 80 percent (4/5), round-trip min/avg/max = 21/31/38 ms

```
Site-1-ASA# show crypto isakmp sa

There are no IKEv1 SAs

IKEv2 SAs:

Session-id:1, Status:UP-ACTIVE, IKE count:1, CHILD count:1

Tunnel-id                 Local                    Remote       Status         Role
  2889995     102.100.100.1/500     102.100.100.2/500       READY     INITIATOR
      Encr: AES-CBC, keysize: 256, Hash: SHA96, DH Grp:5, Auth sign: PSK, Auth verify: PS
K
      Life/Active Time: 86400/552 sec
Child sa: local selector  10.10.11.0/0 - 10.10.11.255/65535
          remote selector 10.10.12.0/0 - 10.10.12.255/65535
          ESP spi in/out: 0x44b661fd/0x206f7ba9


Site-1-ASA# show crypto ipsec sa
interface: outside
    Crypto map tag: outside_map, seq num: 1, local addr: 102.100.100.1

      access-list outside_cryptomap_2 extended permit ip 10.10.11.0 255.255.255.0 10.10.1
2.0 255.255.255.0
      local ident (addr/mask/prot/port): (10.10.11.0/255.255.255.0/0/0)
      remote ident (addr/mask/prot/port): (10.10.12.0/255.255.255.0/0/0)
      current_peer: 102.100.100.2


      #pkts encaps: 4, #pkts encrypt: 4, #pkts digest: 4
      #pkts decaps: 4, #pkts decrypt: 4, #pkts verify: 4
      #pkts compressed: 0, #pkts decompressed: 0
      #pkts not compressed: 4, #pkts comp failed: 0, #pkts decomp failed: 0
      #pre-frag successes: 0, #pre-frag failures: 0, #fragments created: 0
      #PMTUs sent: 0, #PMTUs rcvd: 0, #decapsulated frgs needing reassembly: 0
      #TFC rcvd: 0, #TFC sent: 0
      #Valid ICMP Errors rcvd: 0, #Invalid ICMP Errors rcvd: 0
      #send errors: 0, #recv errors: 0

      local crypto endpt.: 102.100.100.1/500, remote crypto endpt.: 102.100.100.2/500
      path mtu 1500, ipsec overhead 74(44), media mtu 1500
      PMTU time remaining (sec): 0, DF policy: copy-df
      ICMP error validation: disabled, TFC packets: disabled
      current outbound spi: 206F7BA9
```

Introduction to Cisco Firepower Threat Defense (FTD)

Cisco Firepower Threat Defense (FTD) is a next-generation firewall (NGFW) solution that combines traditional firewall features with advanced threat detection and prevention capabilities. It integrates the functionalities of the Cisco Adaptive Security Appliance (ASA) and Sourcefire Firepower, offering a unified platform for security management and policy enforcement

Deployment Options

Hardware Appliances:

Cisco offers dedicated hardware appliances like the Firepower 4100/9300 series, designed for high-performance security deployments.

Software Solutions:

Firepower can also be deployed as a software solution, either on virtual machines or in the cloud.

ASA with FirePOWER Services:

The Cisco ASA (Adaptive Security Appliance) can be enhanced with FirePOWER services to add next-generation firewall capabilities

Management Interfaces:

- **Firepower Management Center (FMC):** A web-based, centralized management platform for managing multiple Firepower devices.
- **Cisco Defense Orchestrator (CDO):** Cloud-based management solution. Ideal for managing a distributed network of FTD devices.
- **Firepower Device Manager (FDM):** A web-based interface for managing individual Firepower devices.
- **Adaptive Security Device Manager (ASDM):** A graphical user interface for managing ASA with FirePOWER Services.

Key Features Firepower

- **Intrusion Detection and Prevention (IDS/IPS):**

  Firepower includes robust intrusion detection and prevention capabilities to identify and block malicious traffic.
- **Advanced Malware Protection (AMP):**

  Cisco AMP helps detect and block advanced malware threats by analyzing files and network traffic.
- **URL Filtering:**

  Firepower allows for granular control over website access, enabling administrators to block or allow specific URLs based on security policies.

- **Application Control:**

  Firepower can identify and control specific applications running on the network, allowing administrators to block or prioritize certain applications.

- **Endpoint Integration:**

  Firepower can be integrated with Cisco endpoint security solutions to provide a holistic security approach.

History of cisco Firewall

Cisco's firewall history began with the PIX (Private Internet Exchange) firewall in 1994, developed by Network Translation Inc. to secure networks. It was later enhanced and integrated into the Adaptive Security Appliance (ASA), which also incorporated features from VPN 3000 and IPS products.

The Firepower Threat Defense (FTD) represents a more recent evolution, combining firewall, intrusion prevention, and advanced threat protection capabilities.

Firepower Hardware Platforms

Cisco Firepower System hardware platforms include physical appliances like the 4100 and 9300 series, which utilize the Firepower extensible Operating System (FXOS), and the 2100 and 1000 series. These platforms can run either Firepower Threat Defense (FTD) or ASA operating systems. The Firepower System also encompasses virtual appliances and cloud-based deployments

Hardware Category	Platform Name/Model Number
Cisco ASA 5500-X Series	ASA5506-X, 5506H-X, 5506W-X, 5508-X, 5516-X, 5512-X, 5515-X, 5525-X, 5545-X, 5555-X
Firepower 2100 Series	Firepower 2110, 2120, 2130, 2140
Firepower 4100 Series	Firepower 4110, 4120, 4140, 4150
Firepower 9000 Series	Firepower 9300

- **4100/9300 Series:**

  These are modular, chassis-based systems designed for high performance and scalability. They support FXOS and can run either FTD or ASA.

- **2100 Series:**

  These platforms also support both FTD and ASA and utilize the FXOS operating system.

- **1000 Series:**

  Similar to the 2100 series, these can run FTD or ASA and use FXOS.

- **ASA 5500-X Series:**

  These older models can also run Firepower Threat Defense software as a module

However, the Cisco Firepower Threat Defense (FTD) unified software cannot be deployed on Cisco ASA 5505 and 5585-X Series appliances. Also note that, as of now Multi Context mode, EIGRP and Multicast is not supported by FTD.

Comparison: Cisco FTD vs Other Firewalls

Feature	Cisco FTD	Palo Alto NGFW	Fortinet FortiGate
Firewall Type	Unified NGFW + IPS + AMP	App-based NGFW	NGFW + UTM
Threat Intelligence	Cisco Talos	WildFire	FortiGuard
Centralized Management	FMC / CDO	Panorama	FortiManager / FortiCloud
Sandboxing	Advanced Malware Protection (AMP)	WildFire	FortiSandbox
Application Control	Yes	Yes	Yes
URL Filtering	Yes	Yes	Yes
SSL Decryption	Yes	Yes	Yes
VPN Support	Site-to-site, Remote Access	Yes	Yes
Ease of Use	Moderate (more enterprise-level)	High	Moderate
Best For	Enterprise & Data Center Security	App-centric environments	SMB to Large Enter

FTD Initial Setup

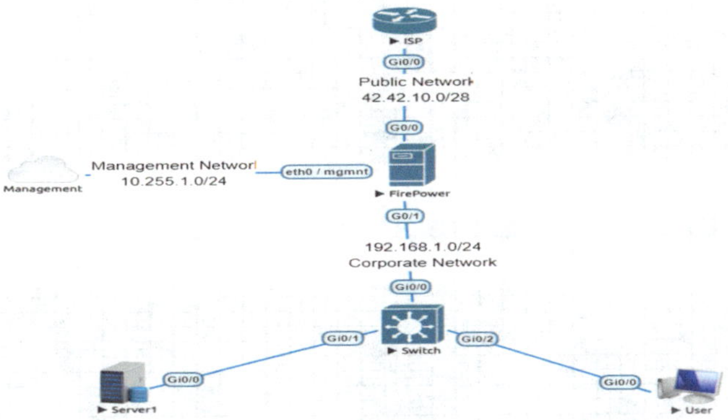

TASK: Apply the initial setup to get the FDM working for management

- Setting up the initial configuration of Cisco FTD is a critical first step to get your system up and running. The setup can be done either via CLI (for basic settings) or GUI (via Firepower Device Manager or Firepower Management Center).
- Here's a **step-by-step guide** for **initial setup of FTD using the CLI**

Access the FTD Console

- Connect to the console via serial (physical device) or virtual console (VM).
- Power on the appliance or start the virtual machine.

Using the CLI for initial setup

Default username: admin, Password: Admin123

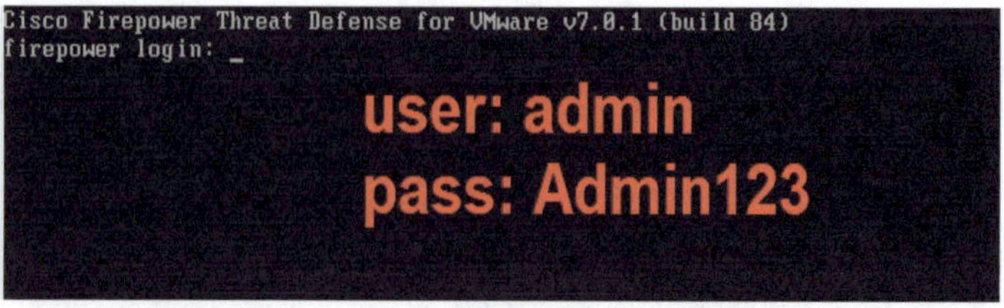

Then the license must be accepted.

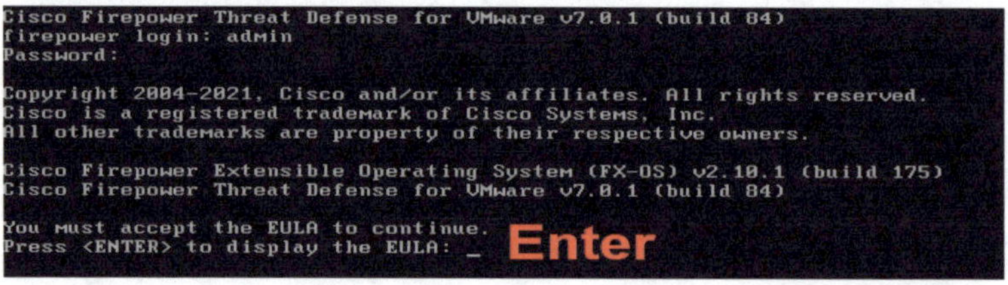

Enter a new password must be configured. This lab we are using nptc123

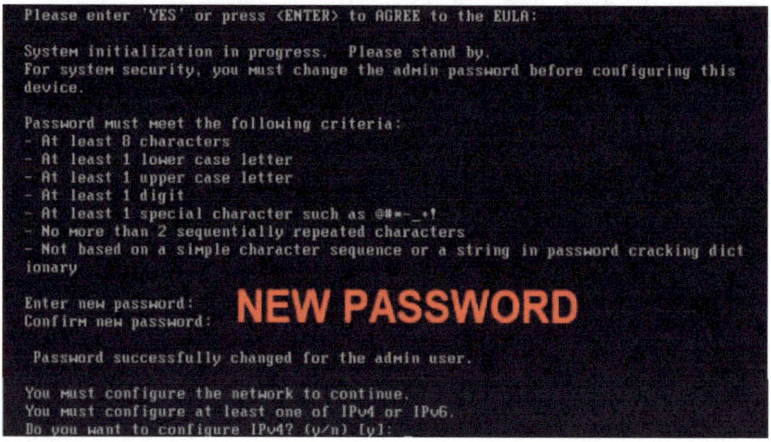

Then IP address, subnet mask, gateway address, the name of FTD, list of DNS servers and domain-name must be configured.

Manage the device locally? Please set it to "no" which means it will be configured through FMD or FMC management server.
Configure **firewall mode**? We will talk about the difference between transparent and routed mode in a special video. But for the beginning, we will start with routed mode. This means that FTD is a part of our routing infrastructure and must take part in our routing topology.

Enter the following information:

Configure IPv4 via DHCP? (y/n) [n]: n

Enter an IPv4 address: 10.255.1.202

Enter the netmask: 255.255.255.0

Enter the gateway: 10.255.1.1

Enter a hostname: FTD-FW 1

Configure DNS servers? (y/n) [y]: y

Enter DNS server 1: 8.8.8.8

With command "**show managers**", we can check if FTD is connected to any FDM manager.

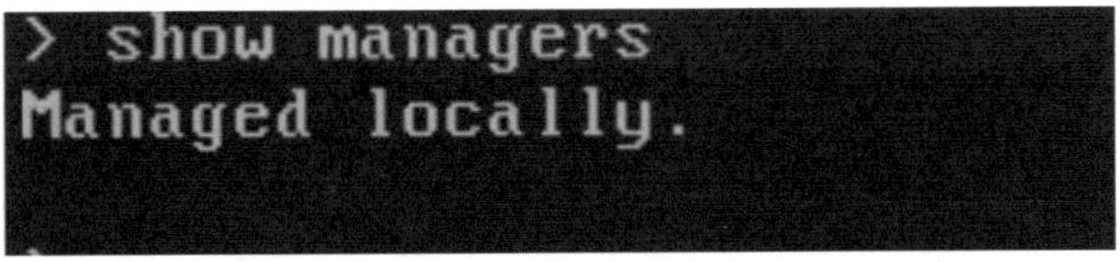

With command **"Show network "**gives you all the network details

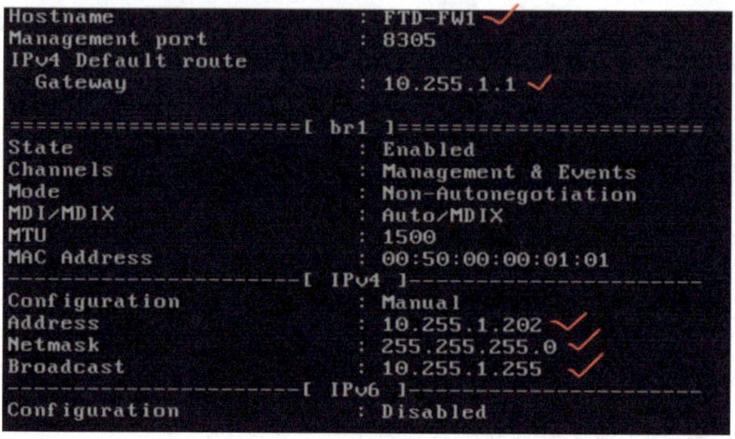

```
Hostname                  : FTD-FW1
Management port           : 8305
IPv4 Default route
  Gateway                 : 10.255.1.1
======================[ br1 ]======================
State                     : Enabled
Channels                  : Management & Events
Mode                      : Non-Autonegotiation
MDI/MDIX                  : Auto/MDIX
MTU                       : 1500
MAC Address               : 00:50:00:00:01:01
----------------------[ IPv4 ]---------------------
Configuration             : Manual
Address                   : 10.255.1.202
Netmask                   : 255.255.255.0
Broadcast                 : 10.255.1.255
----------------------[ IPv6 ]---------------------
Configuration             : Disabled
```

Now we are allowed to connect to FMD through web interface with username and password that we have already configured.

Open a browser.

Go to: https://10.255.1.202

Log in with:

Username: admin

Password: nptc123

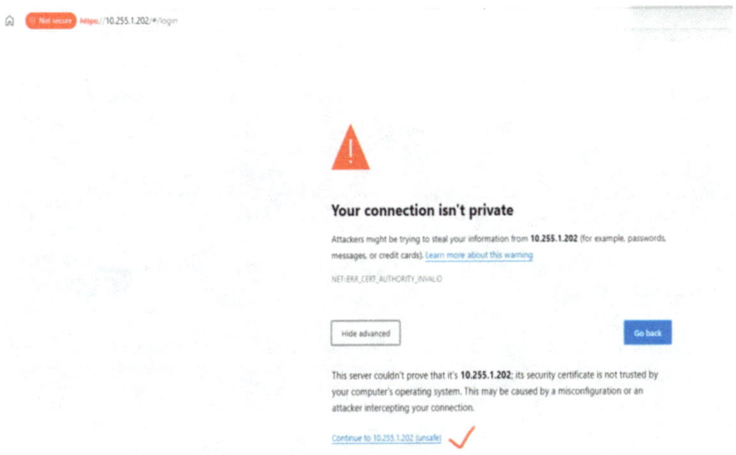
Not secure
https://10.255.1.202/#/login
Your connection isn't private
Attackers might be trying to steal your information from 10.255.1.202 (for example, passwords, messages, or credit cards). Learn more about this warning
NET::ERR_CERT_AUTHORITY_INVALID
Hide advanced
Go back
This server couldn't prove that it's 10.255.1.202; its security certificate is not trusted by your computer's operating system. This may be caused by a misconfiguration or an attacker intercepting your connection.
Continue to 10.255.1.202 (unsafe)

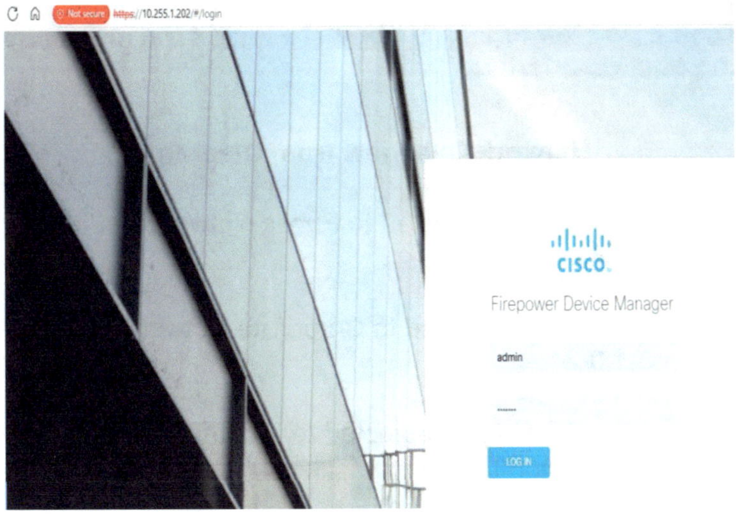
Not secure
https://10.255.1.202/#/login
Firepower Device Manager
admin
LOG IN

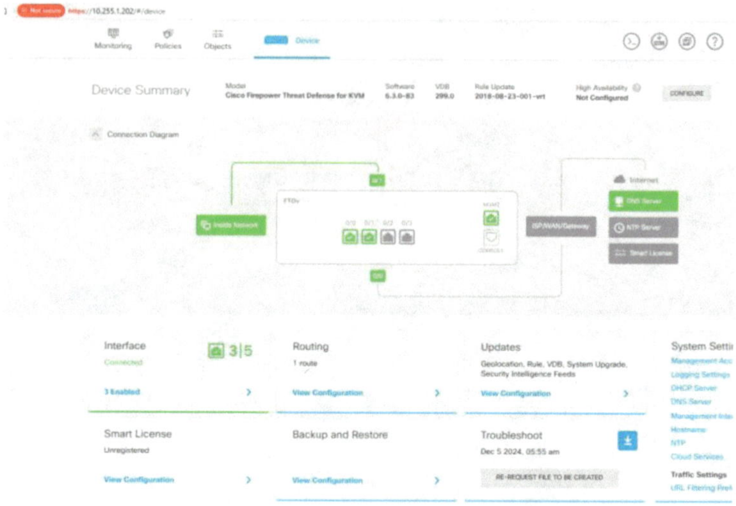

Task 2 Use the topology above to create the inside and the outside interface

Network Overview from Diagram:

Outside (Gi0/0): Connected to Public Network 42.42.10.0/28

Inside (Gi0/1): Connected to Corporate Network 192.168.1.0/24

Management (eth0): Connected to 10.255.1.0/24 (used for FDM access, no policy config here)

Step-1: Navigate to Device > Interfaces

- On the **left menu**, click Device.
- Then go to the **Interfaces** tab at the top.

You'll see a list of all physical interfaces: GigabitEthernet0/0, GigabitEthernet0/1

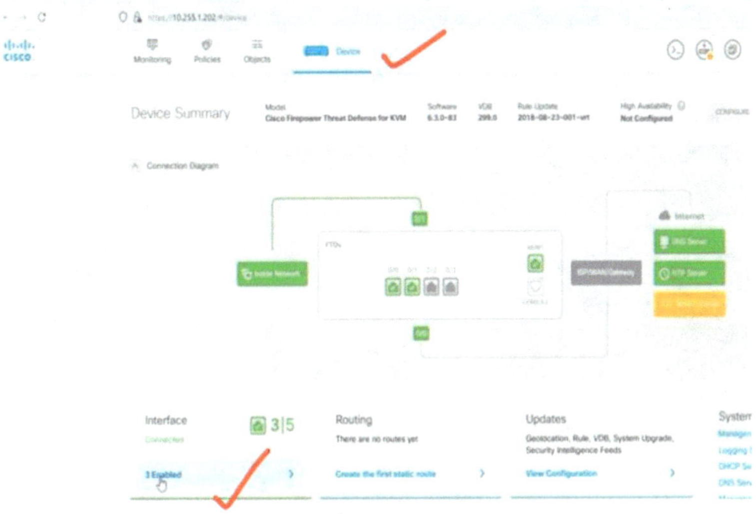

Step 2: Configure Outside Interface (Gi0/0)

1. Click on GigabitEthernet0/0
2. A configuration panel appears on the right.
3. Enable the interface by checking **"Enable this Interface"**
4. Set:
   - **Name:** outside
   - **Security Zone:** Create or select OUTSIDE
   - **IP Configuration:**
     - Select **Static IP**
     - Enter IP Address: 42.42.10.5 (example)
     - Subnet Mask: /28 or 255.255.255.240
   - **IPv4 Gateway:** 42.42.10.1 (ISP Router)

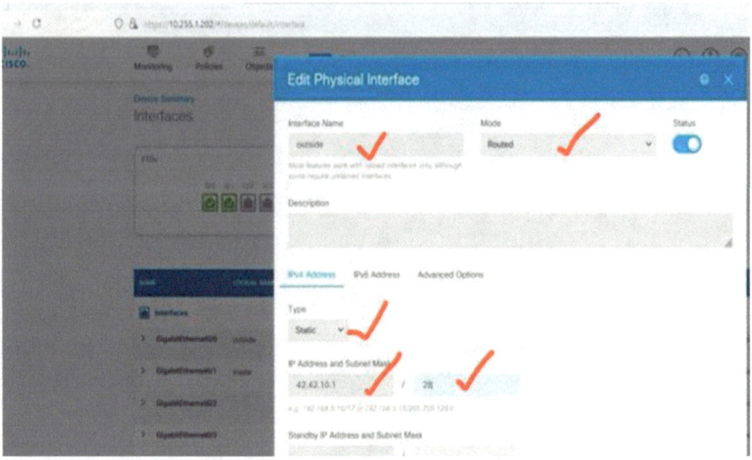

Step 3: Configure Inside Interface (Gi0/1)

1. Click on GigabitEthernet0/1
2. Enable the interface.
3. Set:
    - **Name:** inside
    - **Security Zone:** Create or select INSIDE
    - **IP Configuration:**
        - Static IP: 192.168.1.1
        - Subnet Mask: /24 or 255.255.255.0
4. Click **Save**

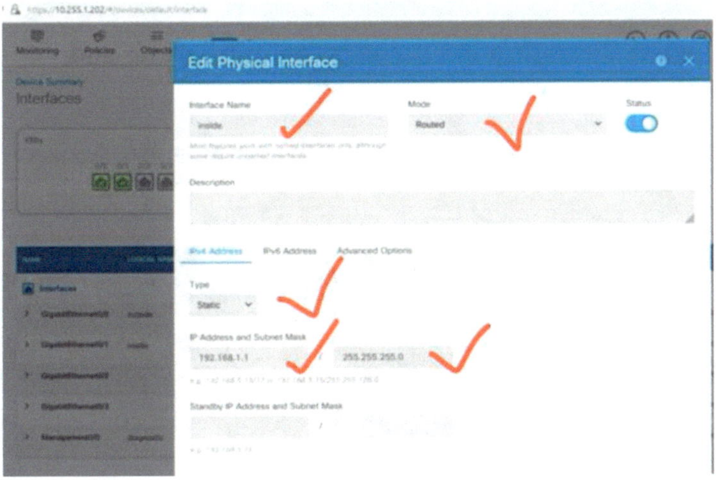

Task3: Changing the Hostname, DNS, NTP

Click on Device > system settings (to make those Changes on hostname, NTP, DNS Server)

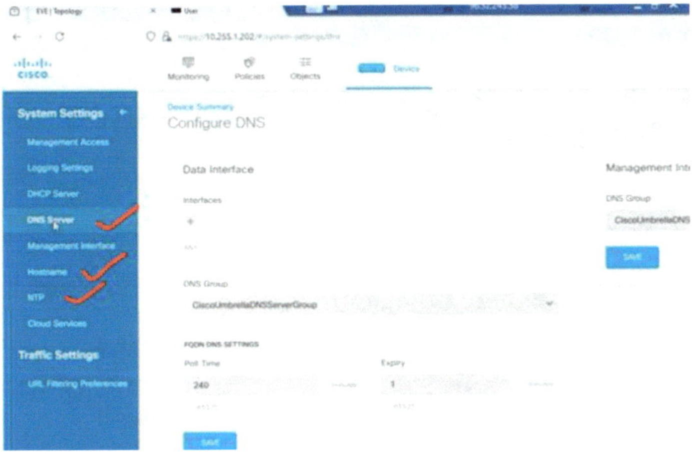

Task 4: Changing the management interface and method

Click on Device > system settings>management Access

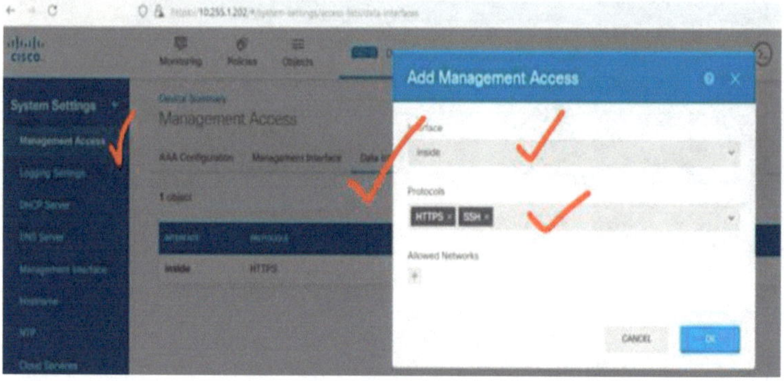

Task 5: Creating the Security Zones

Click on object > Security Zones > pick an interface to configure

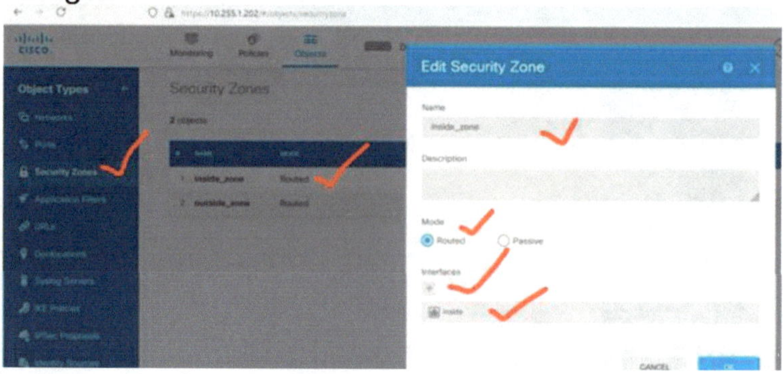

Let's create the outside security zone as well following the same method

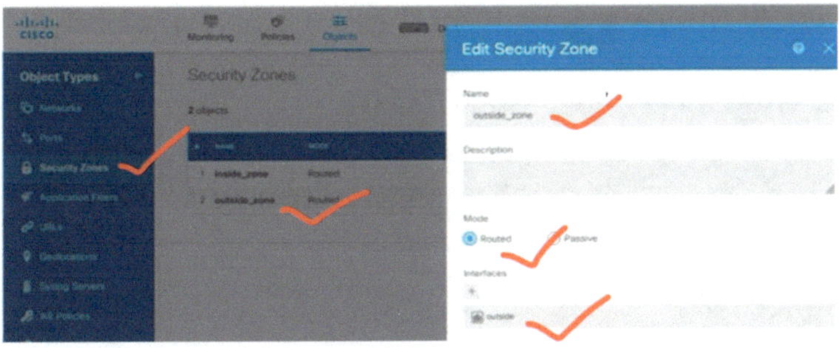

Now the Zones are created

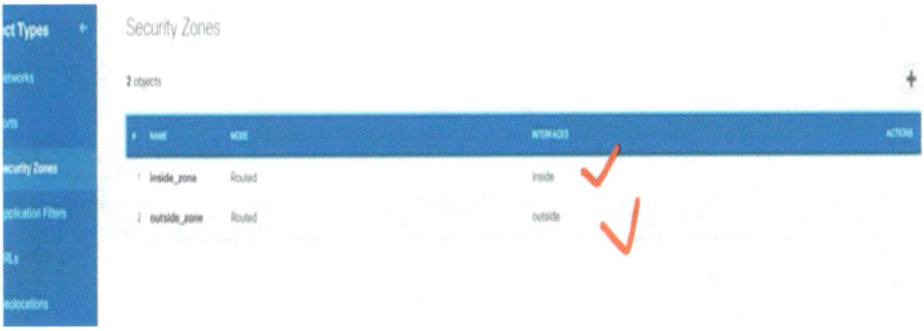

Task 6: Create a default routing

Click device > default static routing

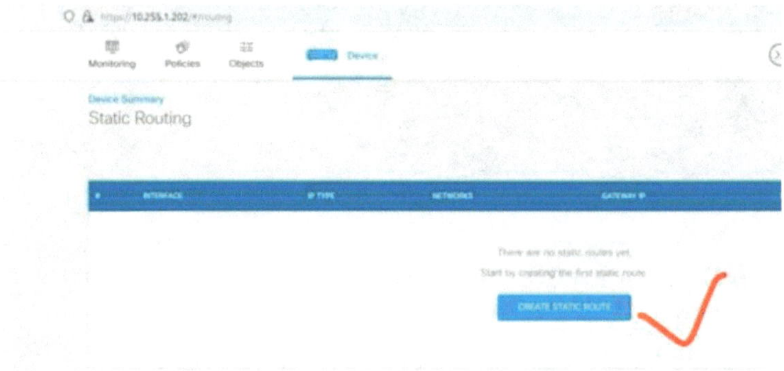

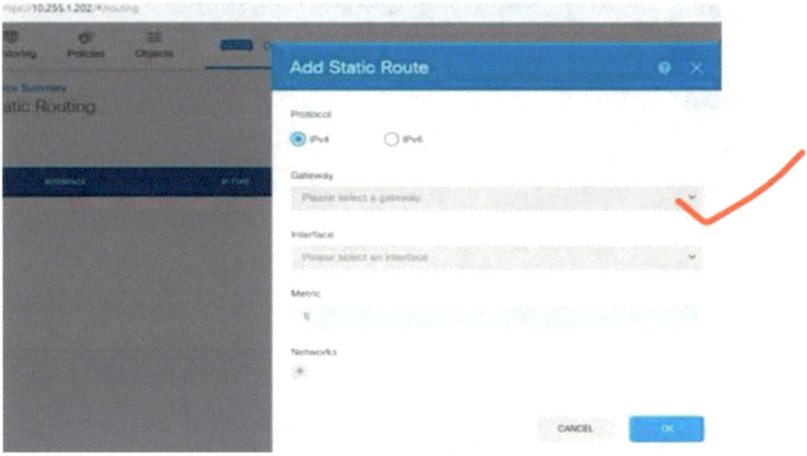

Creating object for the default routing

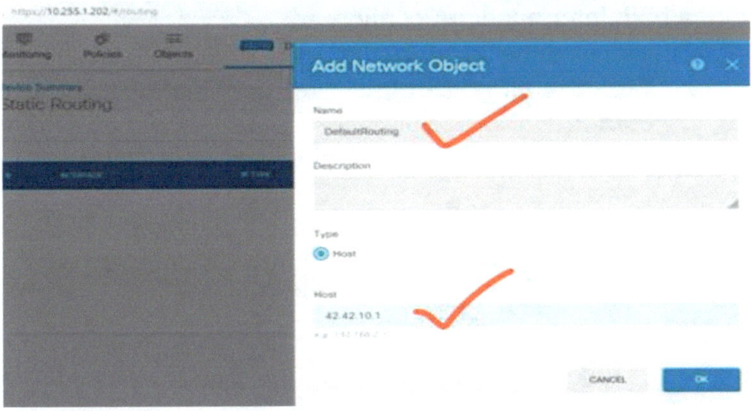

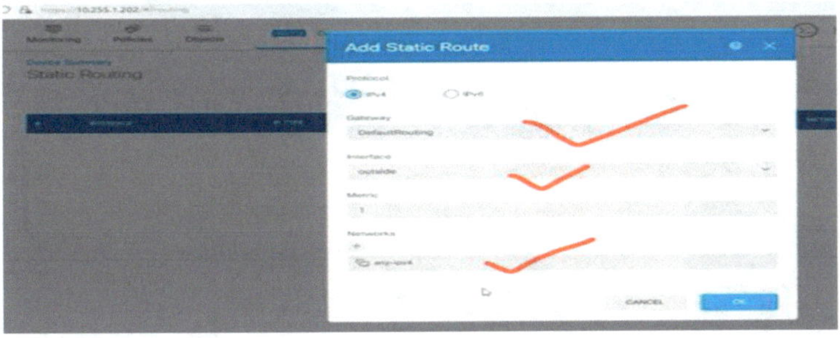

Now we have our default routing

NB: We need to commit the config in order to deploy it

Deploy the Configuration

Once both interfaces are configured:

- Click the **blue "Deploy" button** in the top-right corner.
- Confirm and **deploy the changes** to the FTD.

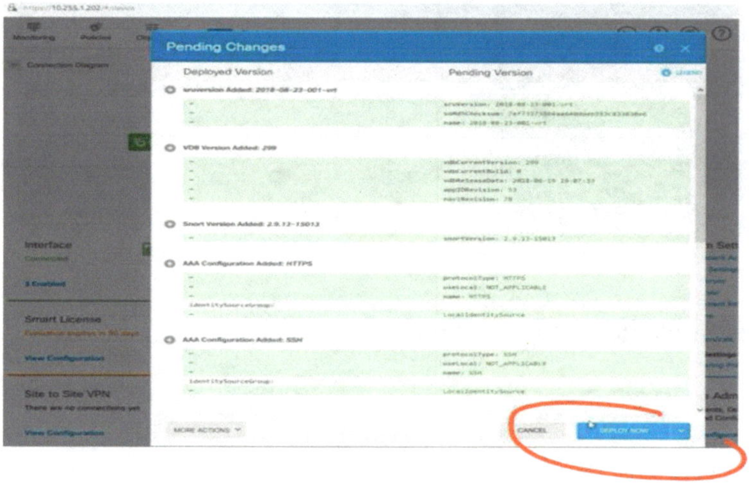

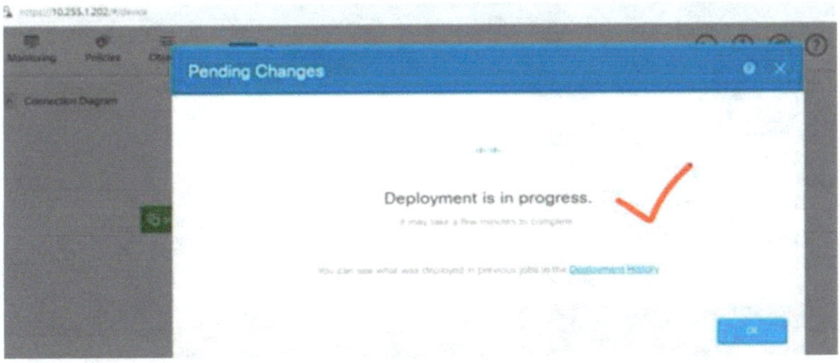

Creating ACL on FTD using FDM

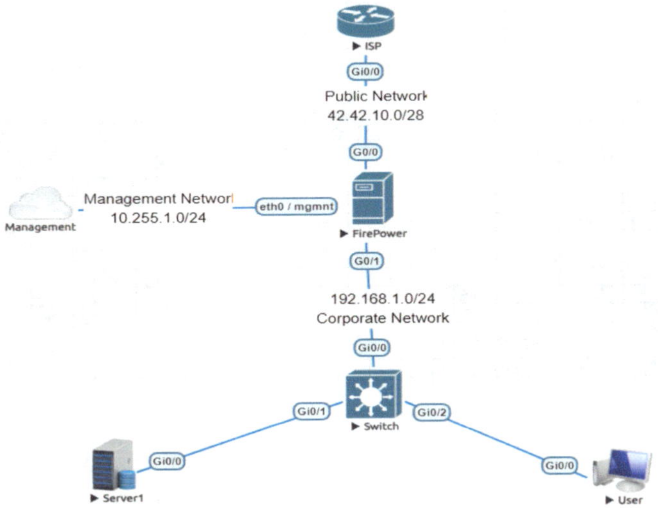

TASK: Creating outbound and Inbound Rules

Step-1: Create your network object

Click on object > network > fill in all the details (for corporate network using 192.168.0.0/24)

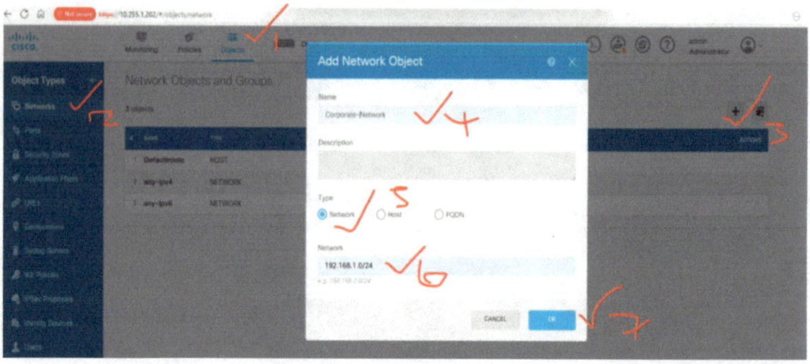

Click on object > network > fill in all the details (for single host ISP using 42.42.10.2)

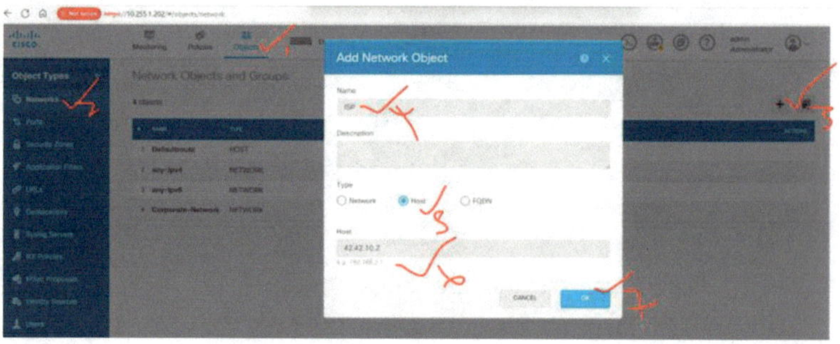

Step 2: Now we can create our outbound and Inbound Rules to allow users on the corporate network have access to the internet.

Click on object > click on the plus sign at far right to add a new rule

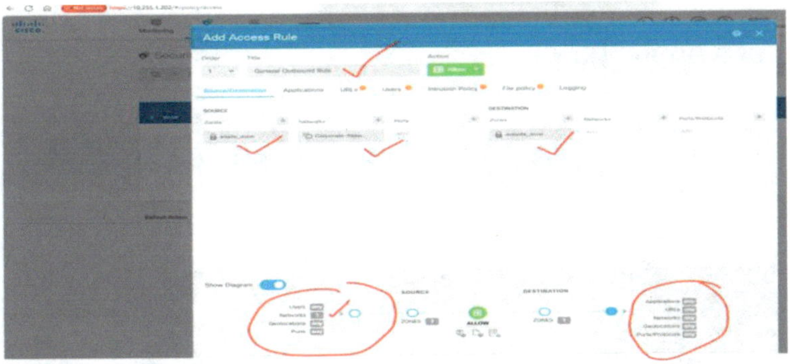

NB: Remember to Deploy

Filter Base on Geographical Location

Task 2; Filter traffic base Geolocation, block traffic to all these countries and continent china, india , russia and Africa as blacklist

Step 1 create object for countries under the black list

Click on objects> geolocations > click on plus sign

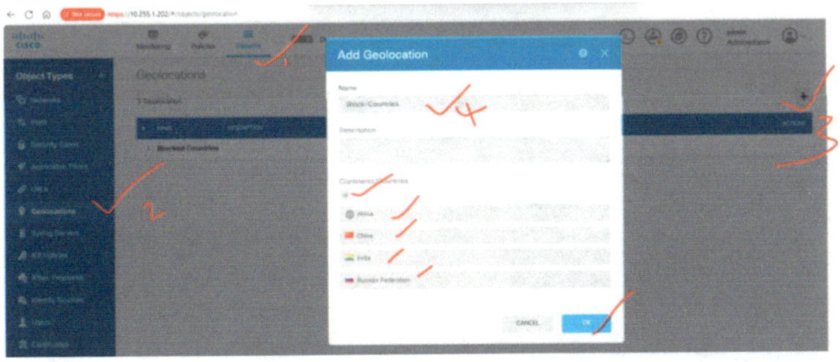

Let’s apply the rule

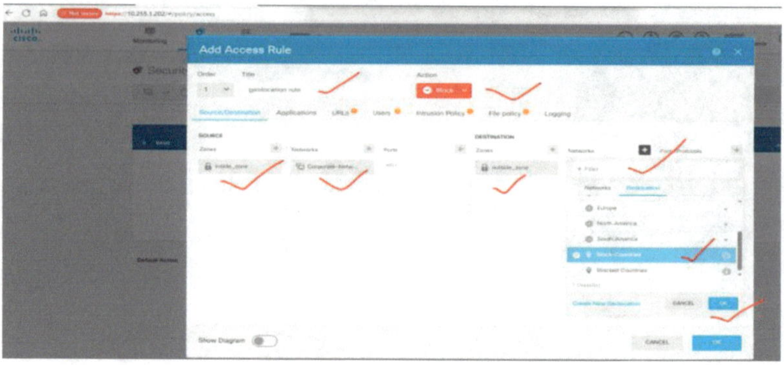

NB: Remember to Deploy

Filter Base on URL

Task 2: Filter base on URL eg facebook, tiktok and inappropriate web nudity and sex

Step 1: Create object for countries under the black list

Click on objects> URL> click on plus sign

For Amazon

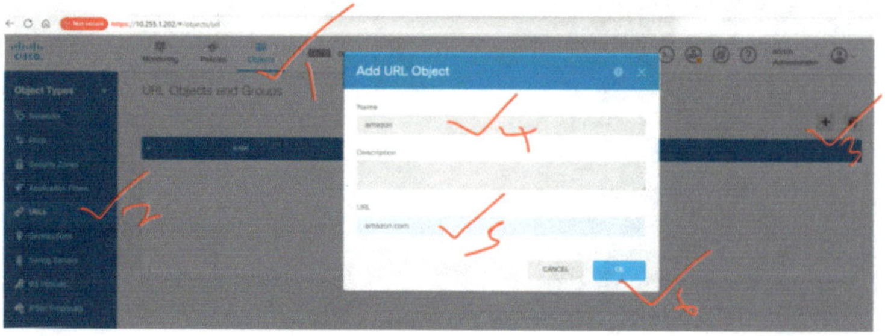

For tiktok

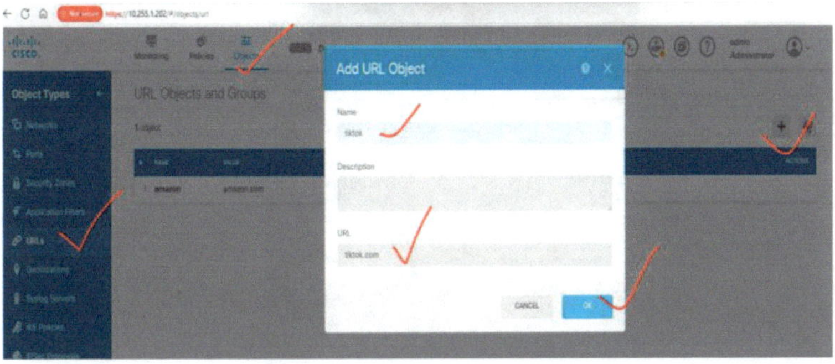

Step 2: Create the group object for both object created

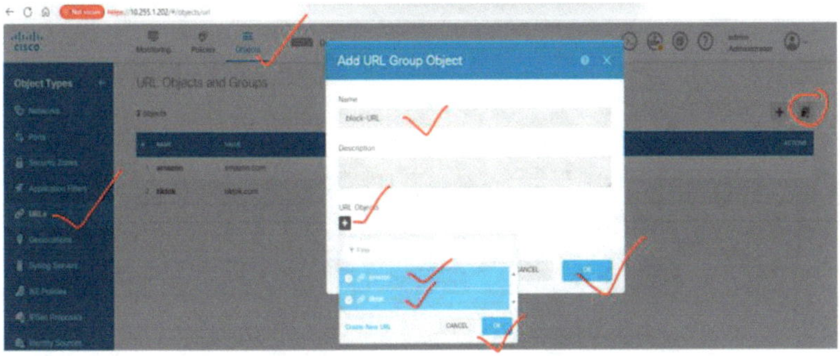

Step 3: Now we can apply ACL

Click policy > click on the plus + sign

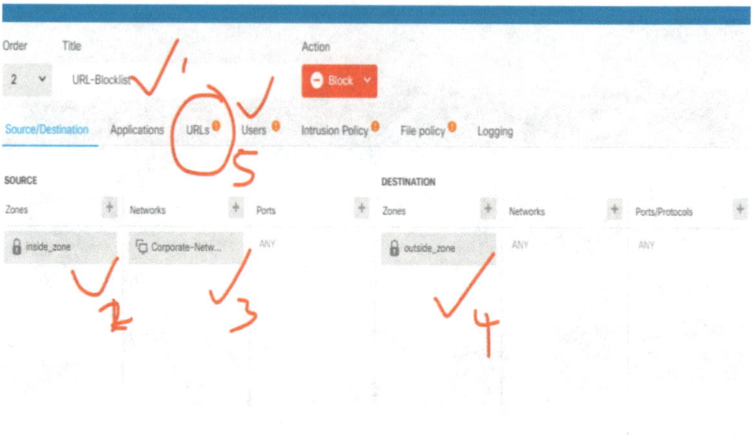

Step 4: Below is the 2-block state and always leave the default Action to trust

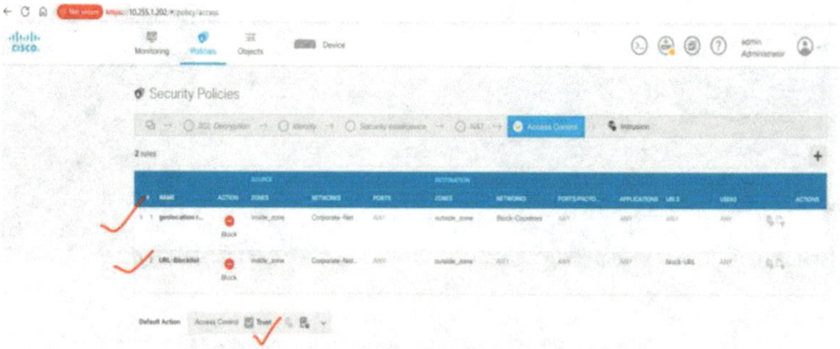

Filter Base on Domain

Task: Filter base on application eg yahoo voice, torrent

Step 1: Apply the ACL no object needed for this

Click policy > click on the plus + sign

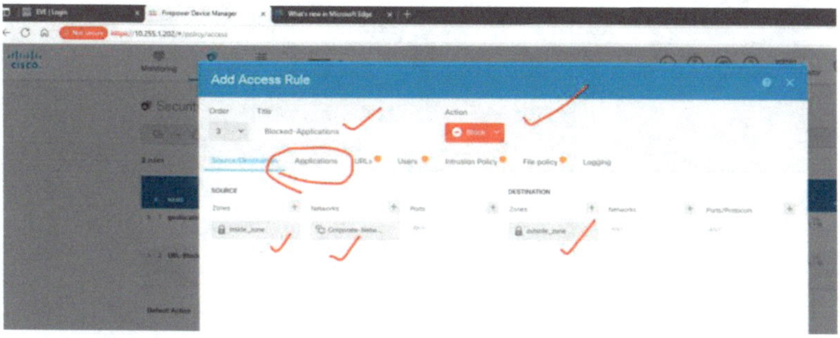

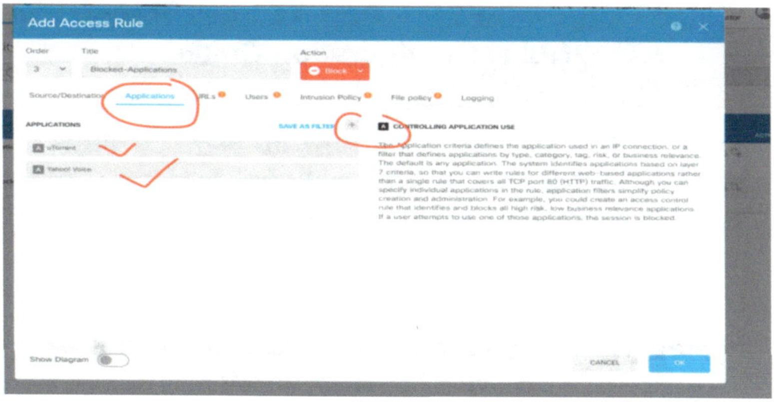

NB: Remember to Deploy

Cisco Firepower FDM – Port Address Translation (PAT)

Project Lab

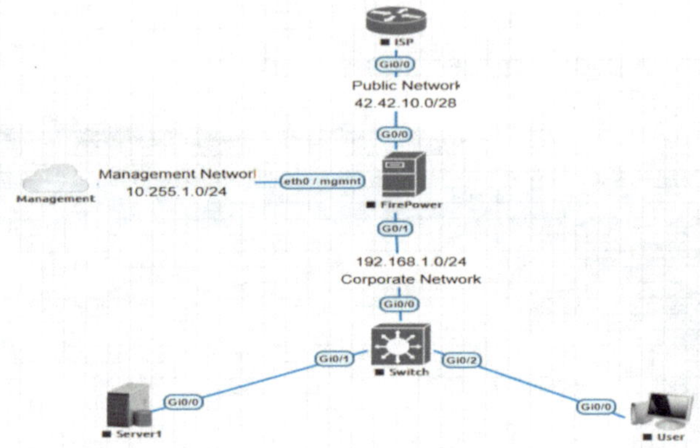

Cisco Firepower FDM – Port Address Translation (PAT) Configuration Guide

Task: This example assumes you want to allow internal users on a private network (e.g., 192.168.1.0/24) to access the internet using the outside interface IP **with dynamic PAT**

Step 1: After your initial setup Log into Firepower Device Manager (FDM)

- Open a web browser and connect to the device's management IP.
- Example: https://192.168.1.1
- Login with admin credentials.

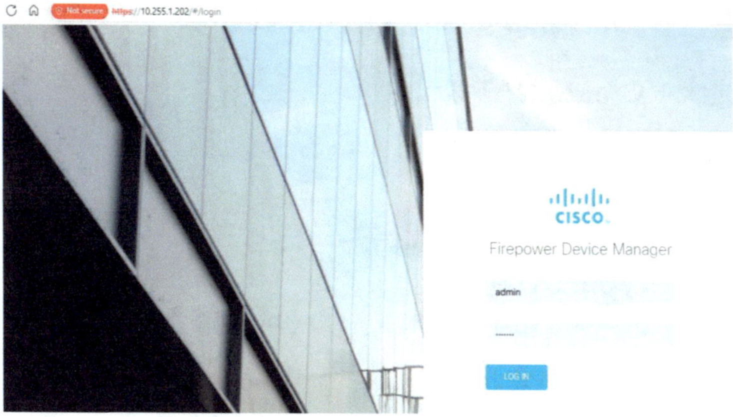

Step-2: Navigate to NAT Configuration

- Go to:
  Policies > NAT

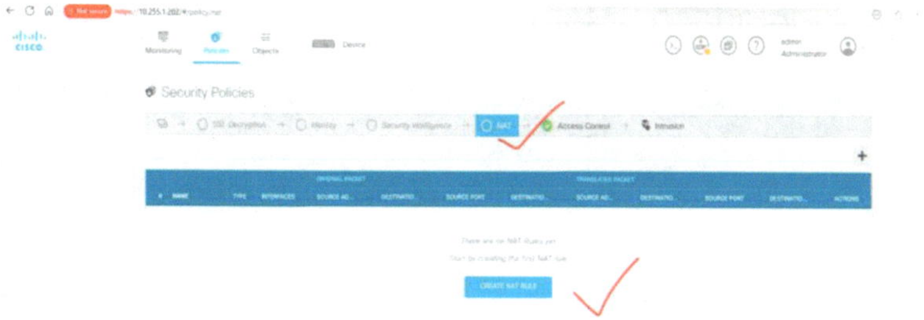

Step-3: Add a New NAT Rule

Click **+ Create NAT Rule**

General:

- **Name**: PAT-to-Internet
- **Type**: Dynamic
- **Source Interface**: inside (or your internal zone)

- **Source Network**: 192.168.1.0/24 (or use a network object in this case Corporate-network)
- **Destination Interface**: outside

- **Translation Type**: Interface PAT
- **Use Interface IP**: ✅ (this will use the outside interface IP for PAT)

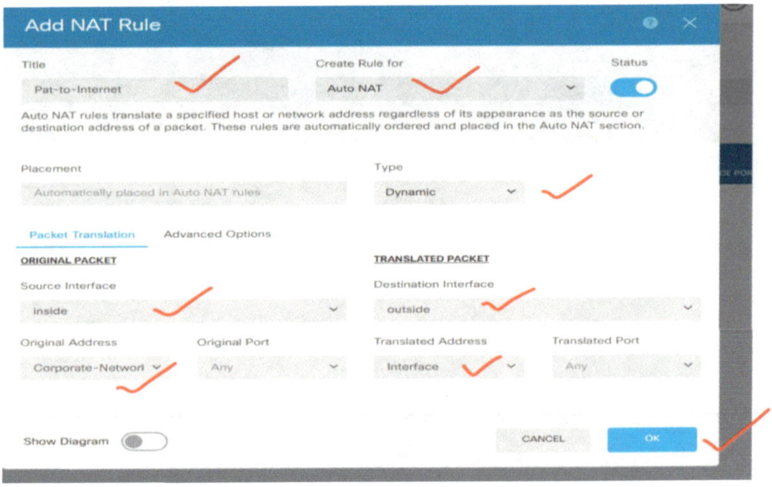

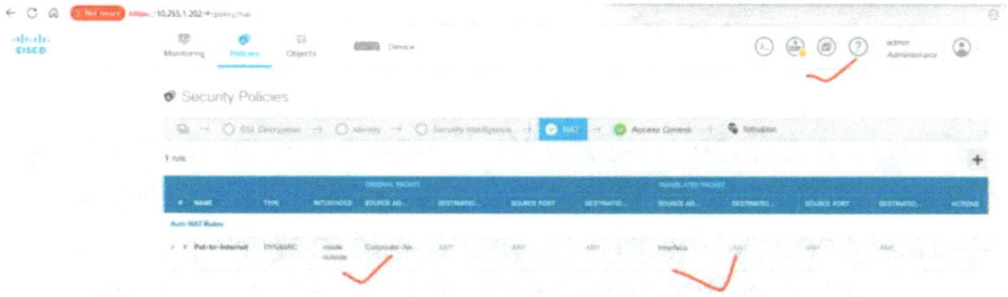

Step 4: Save and Deploy

- Click **Save**
- Then click **Deploy** (top-right corner) to apply the changes to the device.

Site to Site VPN with FTD

Setting up a **Site-to-Site VPN** using **Cisco Firepower Threat Defense (FTD)** allows two remote networks to securely communicate over the internet. Below is a step-by-step guide for configuring it using **Firepower Device Manager (FDM)**.

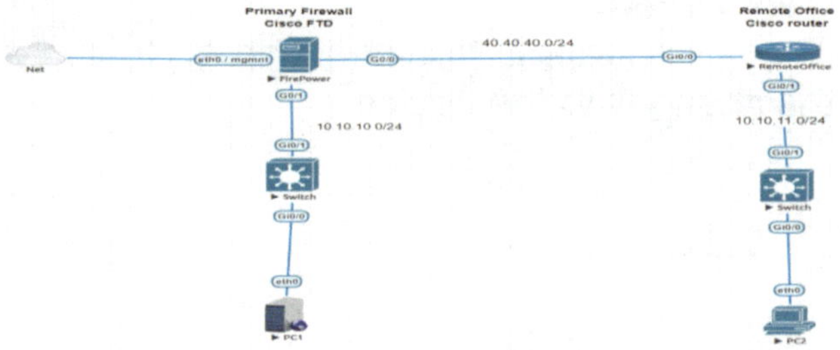

Step-1: Assign IP addresses for both inside and outside network on and FTD and create same on Remote office and ensure the LAN is working both side

Step-2: use default route as your routing protocol

Step-3: create an object for both the corporate network and the remote office network

For Corporate Network

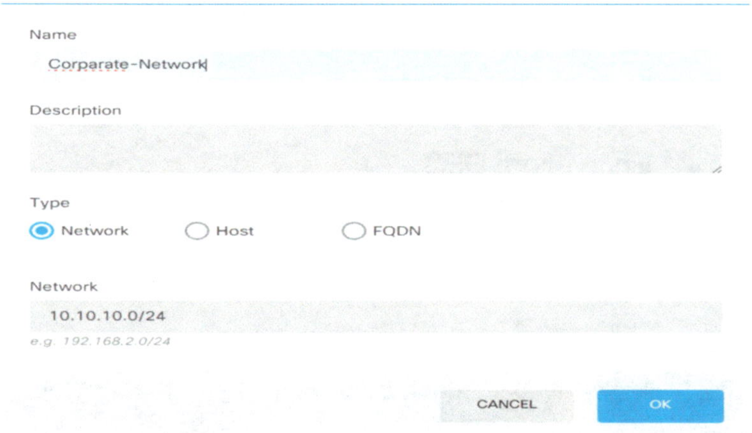

For Remote Office

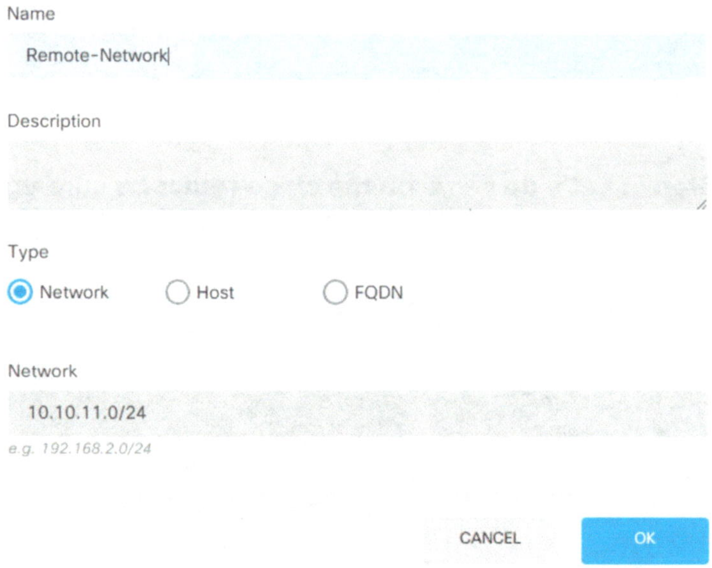

Step: 4 We create our IKE Parameters for Phase 1

KEv2 Settings (Phase 1):

- Encryption: DES
- Integrity: MD5
- DH Group: 2
- Lifetime: 86400 (default)

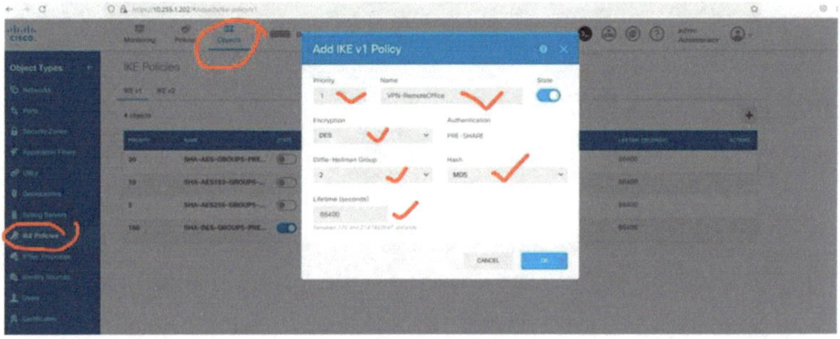

Step 5: Let's do same on the cisco router be sure you have assign your IP address and can ping the FTD

```
RemoteOffice_1#show ip int br
Interface              IP-Address      OK? Method Status                Protocol
GigabitEthernet0/0     40.40.40.2      YES NVRAM  up                    up
GigabitEthernet0/1     10.10.11.1      YES NVRAM  up                    up
GigabitEthernet0/2     unassigned      YES NVRAM  administratively down down
GigabitEthernet0/3     unassigned      YES NVRAM  administratively down down
RemoteOffice_1#ping 40.40.40.1
Type escape sequence to abort.
Sending 5, 100-byte ICMP Echos to 40.40.40.1, timeout is 2 seconds:
!!!!!
Success rate is 100 percent (5/5), round-trip min/avg/max = 3/6/14 ms
RemoteOffice_1#show run | sec route
ip route 0.0.0.0 0.0.0.0 40.40.40.1
```

RemoteOffice_1#conf t
Enter configuration commands, one per line. End with CNTL/Z.
RemoteOffice_1(config)#**crypto isakmp policy 1**
RemoteOffice_1(config-isakmp)#**hash md5**

RemoteOffice_1(config-isakmp)#**authentication pre-share**
RemoteOffice_1(config-isakmp)#**group 2**
RemoteOffice_1(config-isakmp)#**lifetime 86400**
RemoteOffice_1(config-isakmp)#

Step 6: We create Parameters for Phase 2 (IPsec Policy)

Encryption: ESP

- Integrity: ESP-MD5-HMAC
- PFS: Group 14
- Lifetime: 3600

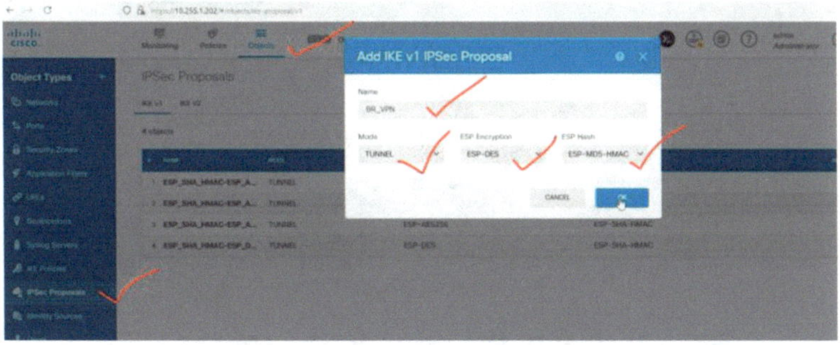

Step-7: Apply same configuration policy on cisco router

RemoteOffice_1 (config)#crypto IPsec transform-set BR2HQ **esp-des esp-md5-hmac**

Step-8: Let's now create the IPsec site-to-site with the pre-share key of Cisco on the FTD

Click on Device > Site-to-Site VPN

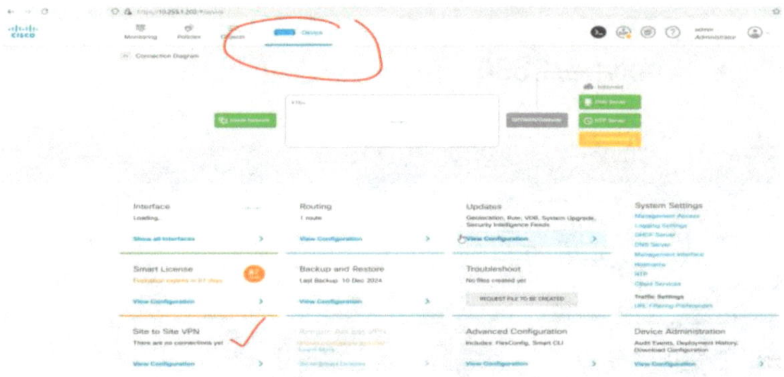

Step-9: Create the connection for the VPN

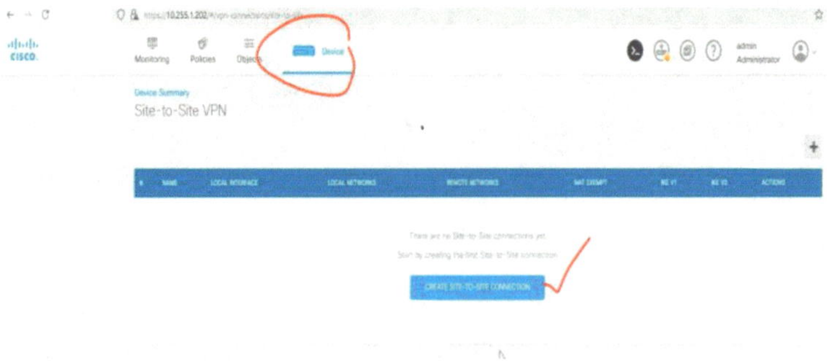

Step-10: Apply the info as indicated below

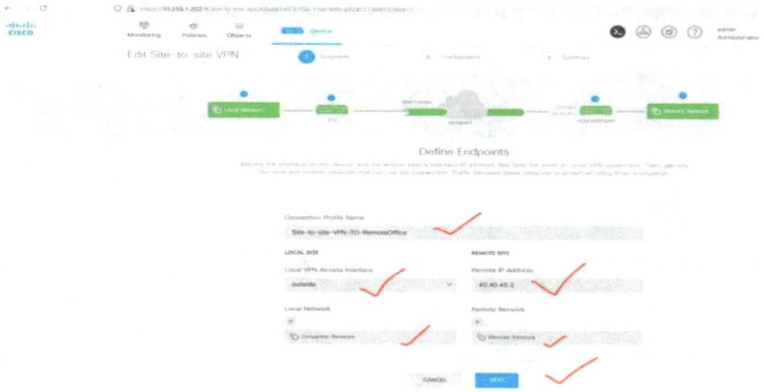

Step-11: Select the IKE (Phase-1) you created as object

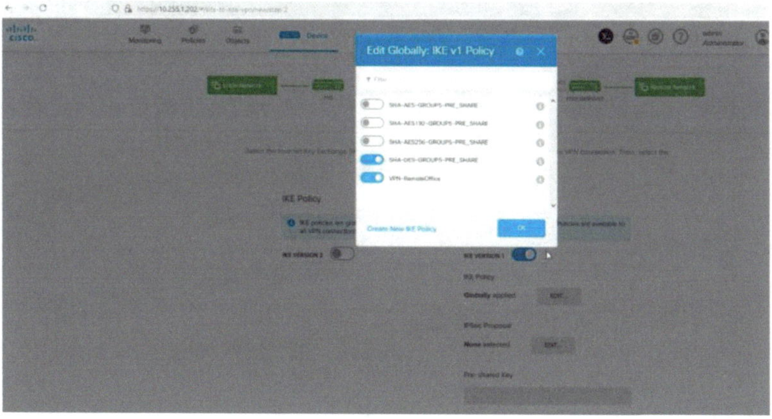

Step-12: Select the IPsec Proposal for Phase 2

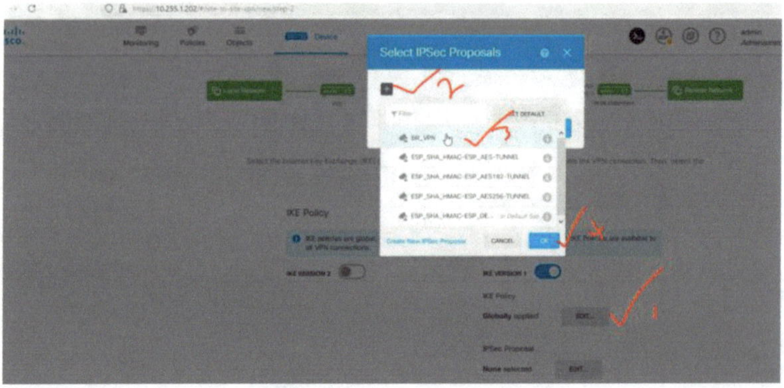

Step-13: Apply on the connection as indicated below

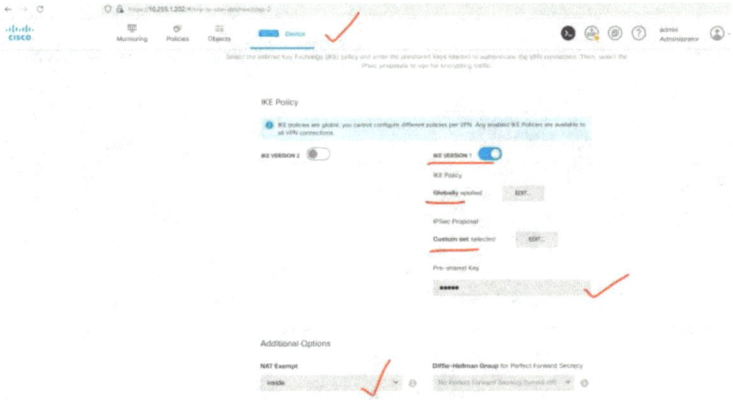

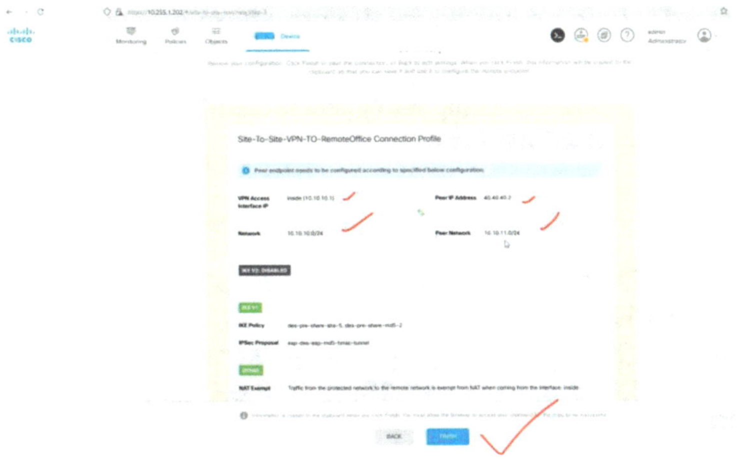

Now the connection is established as indicated below

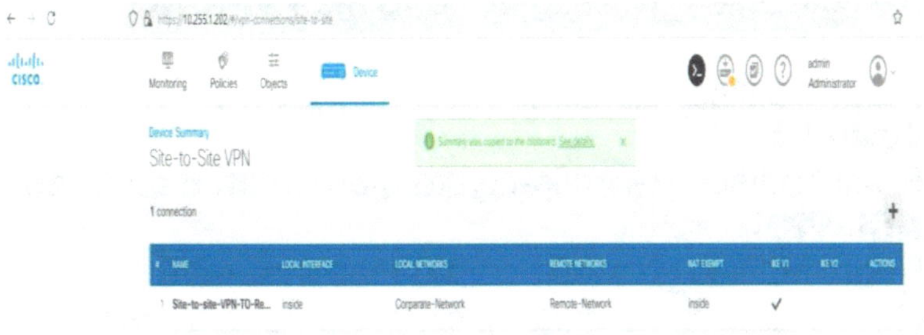

Step-14: let's no configure the pre-share on Cisco router

RemoteOffice_1(config)#**crypto isakmp key cisco address 40.40.40.1**

Step-15: Create ACL for the VPN

RemoteOffice_1(config)#**ip access-list extended ACL-VPN**
RemoteOffice_1(config-ext-nacl)#**permit ip 10.10.11.0 0.0.0.255 10.10.10.0 0.0.0.255**

Step-16: Let's Configure the crypto map policy

RemoteOffice_1(config-ext-nacl)#
RemoteOffice_1(config-ext-nacl)#exit
RemoteOffice_1(config)#**crypto map VPN-HQ 1 ipsec-isakmp**
RemoteOffice_1(config-crypto-map)#**set peer 40.40.40.1**
RemoteOffice_1(config-crypto-map)#**match address ACL-VPN**
RemoteOffice_1(config-crypto-map)#**int g0/0**
RemoteOffice_1(config-if)#**crypto map VPN-HQ**

Step 17: Let's now generate traffic to bring the tunnel up

VPCS> ping 10.10.11.2

10.10.11.2 icmp_seq=1 timeout
84 bytes from 10.10.11.2 icmp_seq=2 ttl=63 time=73.455 ms
84 bytes from 10.10.11.2 icmp_seq=3 ttl=63 time=49.090 ms

84 bytes from 10.10.11.2 icmp_seq=4 ttl=63 time=63.237 ms
84 bytes from 10.10.11.2 icmp_seq=5 ttl=63 time=63.629 ms

Step-18: On Cisco Router

RemoteOffice_1#**show crypto isakmp sa**

```
dst            src            state
conn-id status
40.40.40.2     40.40.40.1     QM_IDLE
1001 ACTIVE
```

RemoteOffice_1#**show crypto ipsec sa**

```
interface: GigabitEthernet0/0
    Crypto map tag: VPN-HQ, local addr 40.40.40.2

   protected vrf: (none)
   local  ident (addr/mask/prot/port):
(10.10.11.0/255.255.255.0/0/0)
   remote ident (addr/mask/prot/port):
(10.10.10.0/255.255.255.0/0/0)
   current_peer 40.40.40.1 port 500
     PERMIT, flags={origin_is_acl,}
    #pkts encaps: 4, #pkts encrypt: 4, #pkts digest: 4
    #pkts decaps: 4, #pkts decrypt: 4, #pkts verify: 4
    #pkts compressed: 0, #pkts decompressed: 0
    #pkts not compressed: 0, #pkts compr. failed: 0
    #pkts not decompressed: 0, #pkts decompress
failed: 0
    #send errors 0, #recv errors 0
```

On FTD

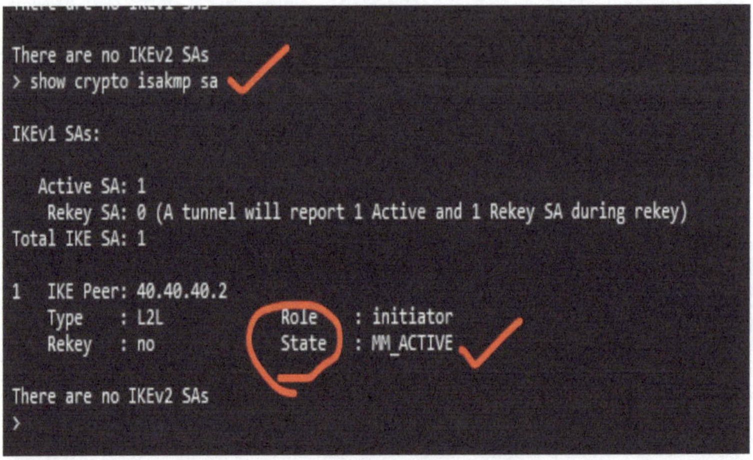

```
There are no IKEv2 SAs
> show crypto isakmp sa

IKEv1 SAs:

   Active SA: 1
    Rekey SA: 0 (A tunnel will report 1 Active and 1 Rekey SA during rekey)
Total IKE SA: 1

1   IKE Peer: 40.40.40.2
    Type    : L2L             Role    : initiator
    Rekey   : no              State   : MM_ACTIVE

There are no IKEv2 SAs
>
```

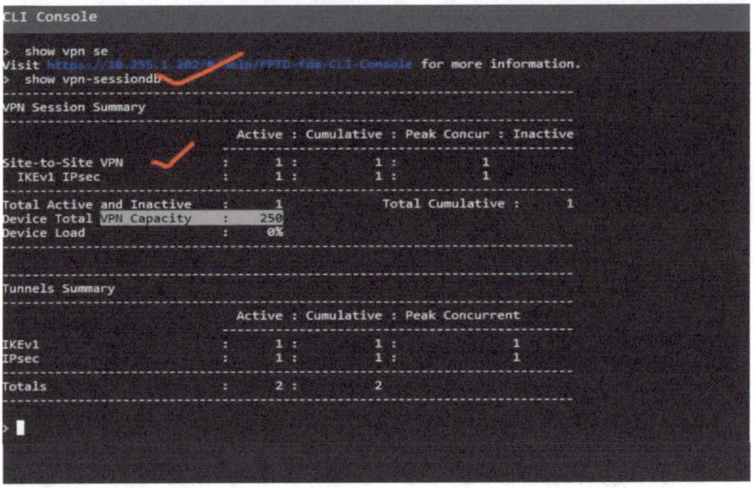

```
CLI Console
>  show vpn se
Visit https://10.255.1.202/#/help/FPTD-fdm-CLI-Console for more information.
>  show vpn-sessiondb
---------------------------------------------------------------------------
VPN Session Summary
---------------------------------------------------------------------------
                              Active : Cumulative : Peak Concur : Inactive
                           ------------------------------------------------
Site-to-Site VPN             :      1 :          1 :           1
  IKEv1 IPsec                :      1 :          1 :           1
---------------------------------------------------------------------------
Total Active and Inactive    :      1           Total Cumulative :       1
Device Total VPN Capacity    :    250
Device Load                  :     0%
---------------------------------------------------------------------------

---------------------------------------------------------------------------
Tunnels Summary
---------------------------------------------------------------------------
                              Active : Cumulative : Peak Concurrent
                           ------------------------------------------------
IKEv1                        :      1 :          1 :             1
IPsec                        :      1 :          1 :             1
---------------------------------------------------------------------------
Totals                       :      2 :          2
---------------------------------------------------------------------------

>
```

Firepower Threat Defense High Availability

Cisco, **FTD HA** stands for **Firepower Threat Defense High Availability**. It's a deployment mode where two Cisco FTD devices (usually firewalls) are paired to provide redundancy and minimize downtime.

FTD HA (High Availability) allows two FTD devices to operate as a failover pair. If one fails, the other takes over seamlessly, ensuring continuous security and connectivity

Key Features of FTD HA

Feature	Description
Failover Types	Active/Standby (only one device handles traffic at a time)
Configuration Sync	Automatically syncs most settings (policies, objects, routes, NAT, etc.)
Link Monitoring	Monitors interfaces and units for failures
Stateful Failover	Optionally maintains session state across failovers (for minimal session interruption)
Health Monitoring	Tracks system, interface, and peer health
Device Management	Usually managed through **Cisco FMC (Firepower Management Center)**

Pre-requisites for FTD HA

Before configuring HA, make sure:

Both FTDs:

- Are the **same model**
- Are running the **same FTD software version**
- Have **factory default config** or minimal setup
- Have at least one **dedicated physical interface for HA link**
- Have **same licensing** (Threat, URL, Malware, etc
-

Connectivity:

- Devices are **cabled directly** or via a switch for the **HA link**
- Both are **reachable via their management IPs**

FTD HA Project

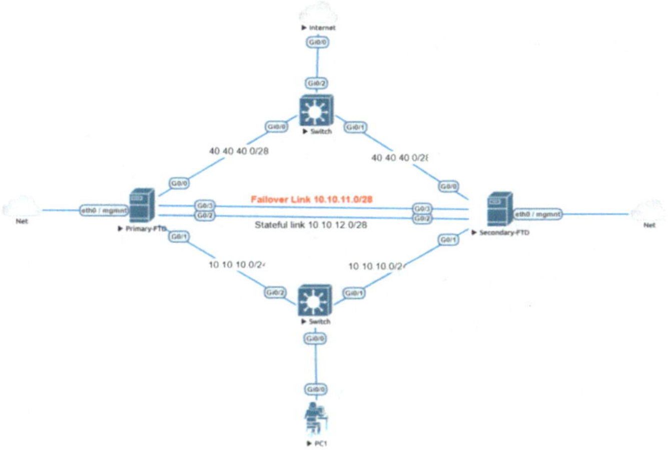

Task: Set up HA using FDM

Step-1 We will start with the Primary FW with the initial set up to get FDM management IP working

Step-2 Configure the outside IP address with the failover I

Click on device> interface > click on port g0/0

Step-3: Configure the inside network

Click on device> interface > click on port g0/1

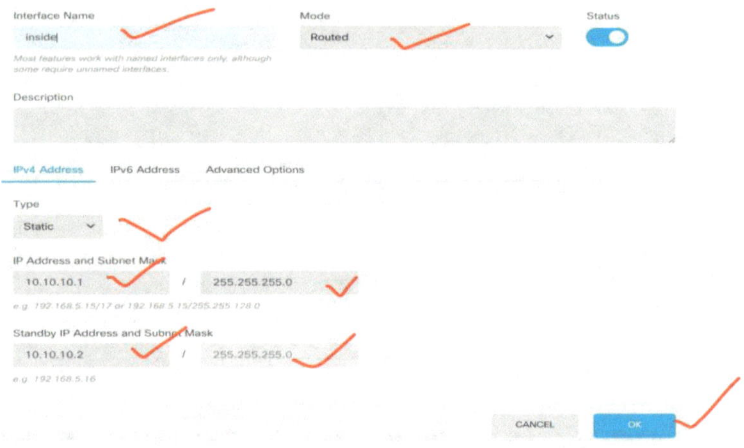

Apply them on Security Zones

INSIDE ZONES

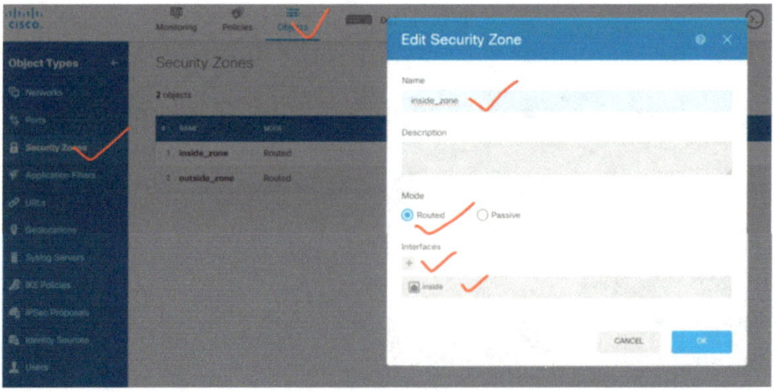

OUTSIDE ZONES

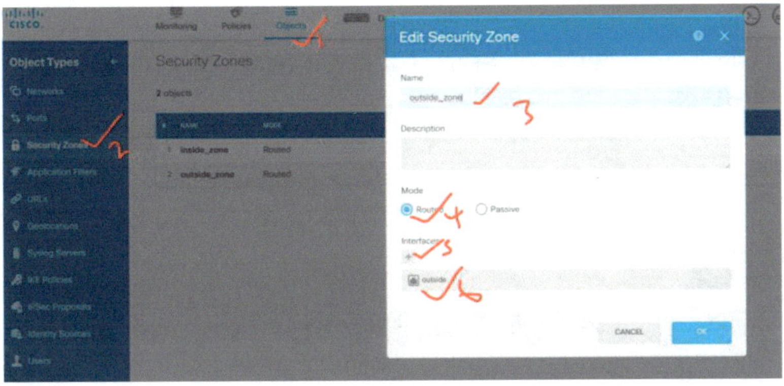

Step-4 Lets Create object for default routing using the your Internet IP on the topology

Click on object>networks>add your object details

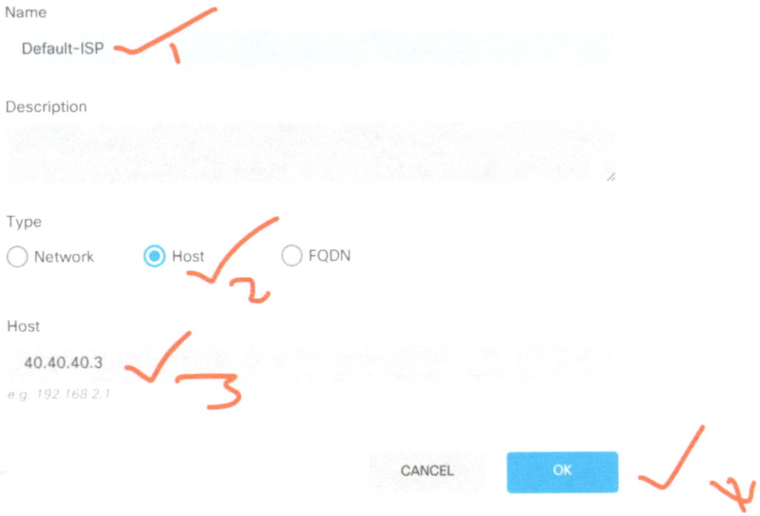

Step-5: Configure your default route

Click on device >Static Routing> add your route details

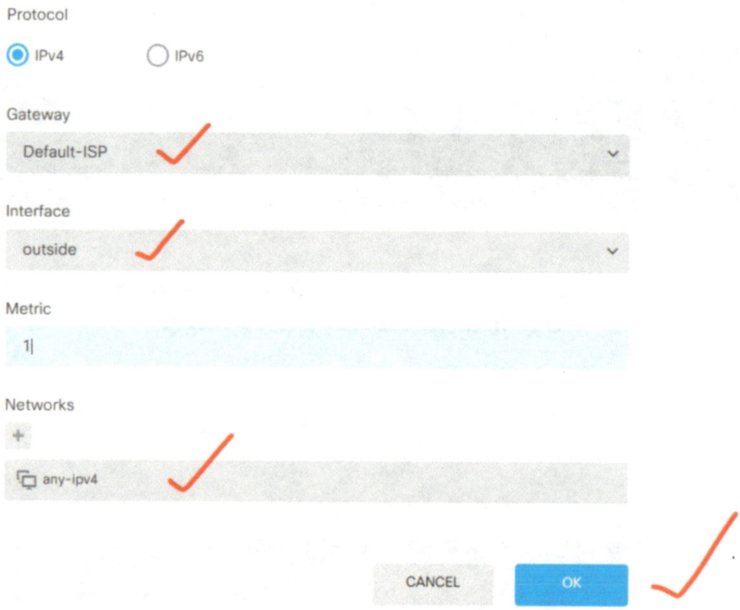

Step 6: Create your inbound policies to allow your inside network to communicate the internet with icmp protocol

Click Policies> add your rules with below details

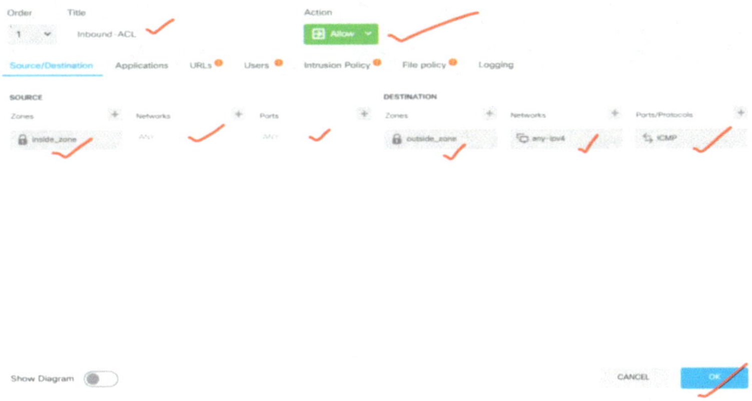

Step 7: let's now set up our failover configuration

Enable the failover links g0/2 and g0/3, remove any default IP addresses

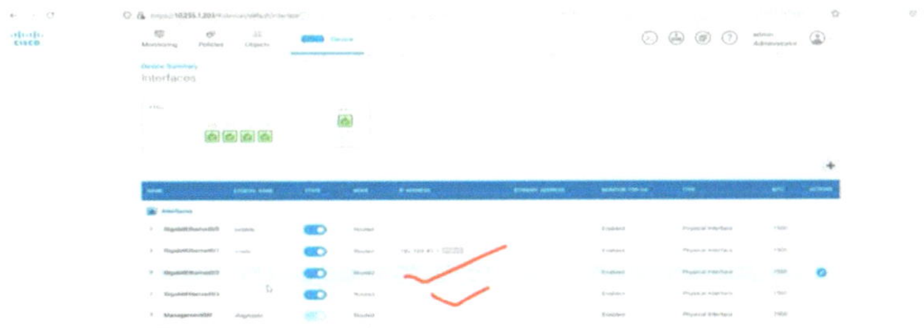

Click on Device > click on configure at high Availability

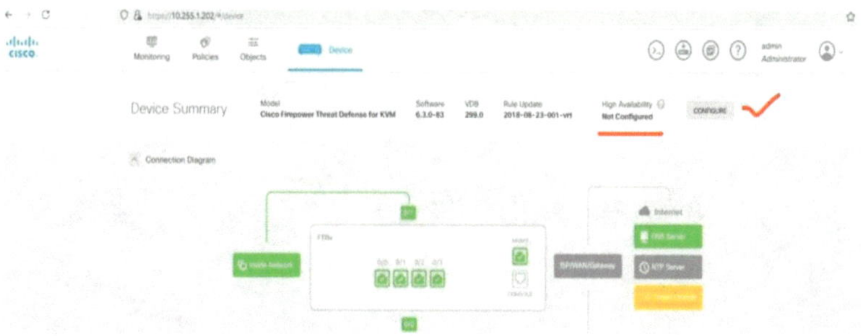

Click on Device > click on configure at high Availability and add all details > activate

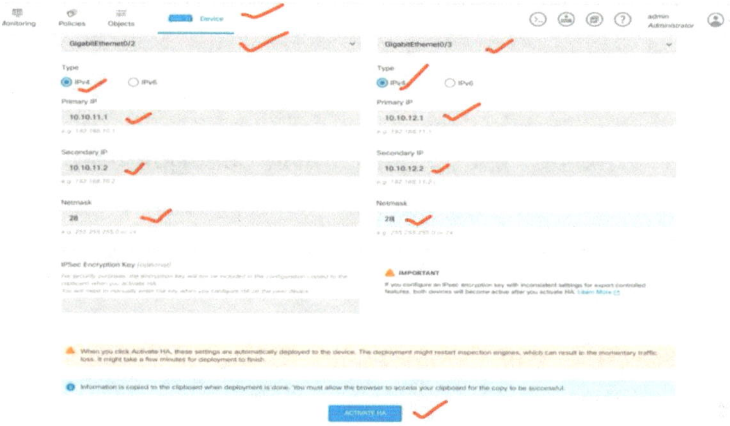

Save and Deploy

- Click **Save**
- Then click **Deploy** (top-right corner) to apply the changes to the device.

Now the Active is ready

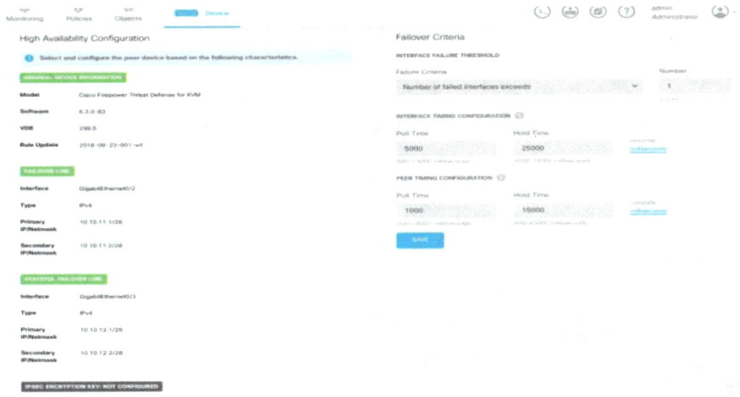

Step 8: We will now configure the Secondary device

Step-9: We will start with the Secondary FW with the initial set up to get FDM management IP and log on

Step-10; Configure the outside IP address with the failover IP

Click on device> interface > click on port g0/0

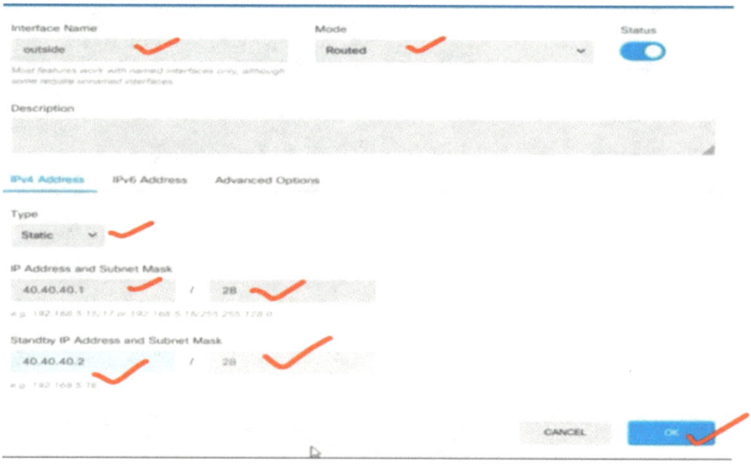

Step-11: Configure the inside network

Click on device> interface > click on port g0/1

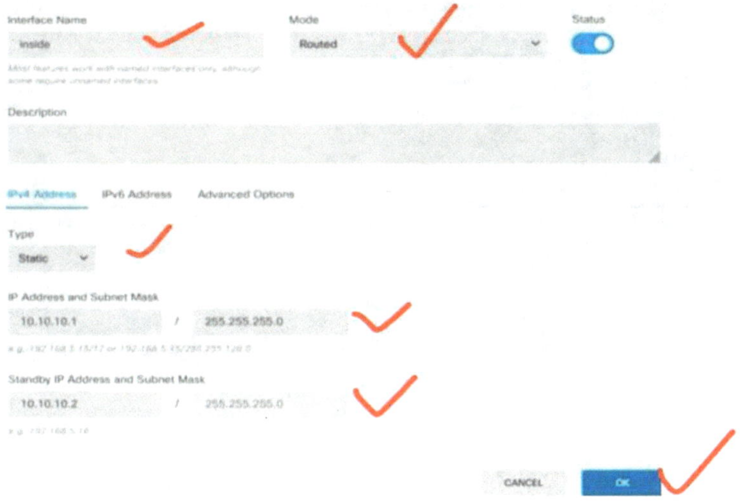

Apply them on security Zones

INSIDE ZONES

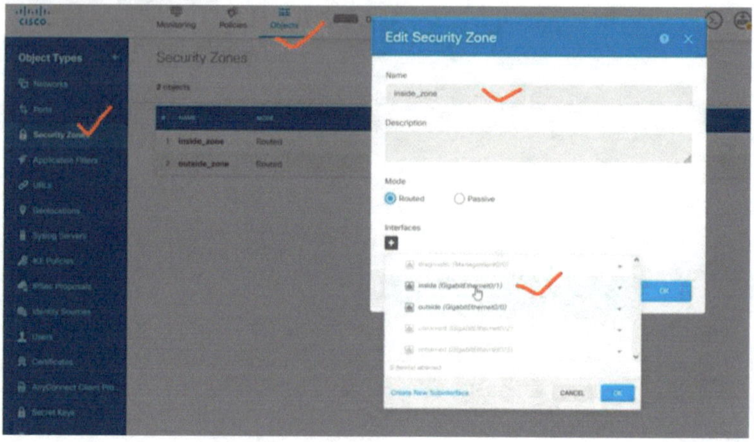

OUTSIDE ZONES

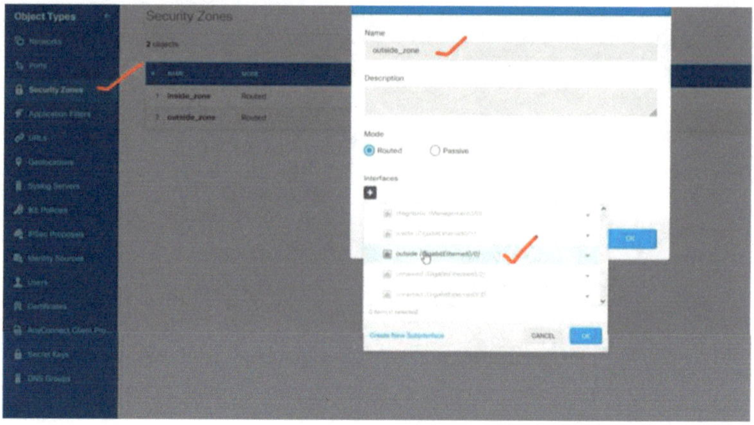

Step-12: Let's do the failover config on the Secondary device using FDM

Click on Device > click on Configure the HA > choose Secondary

Choose Secondary and add all details using the topology

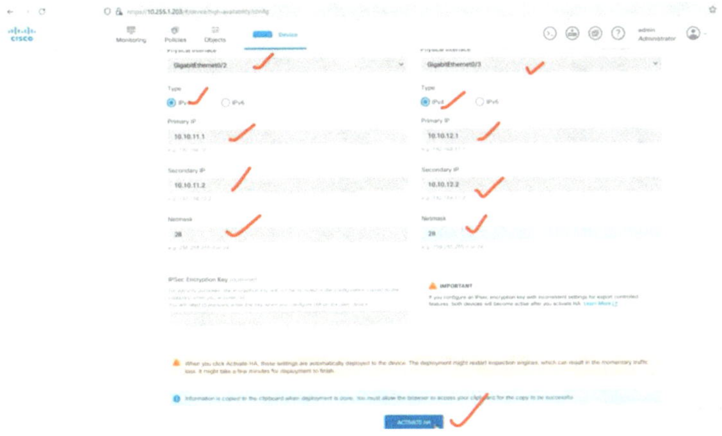

Step-13: Let’s refresh the page to see if the HA is working

This state is a good sign if you see Syncing (Devices will sync configurations)

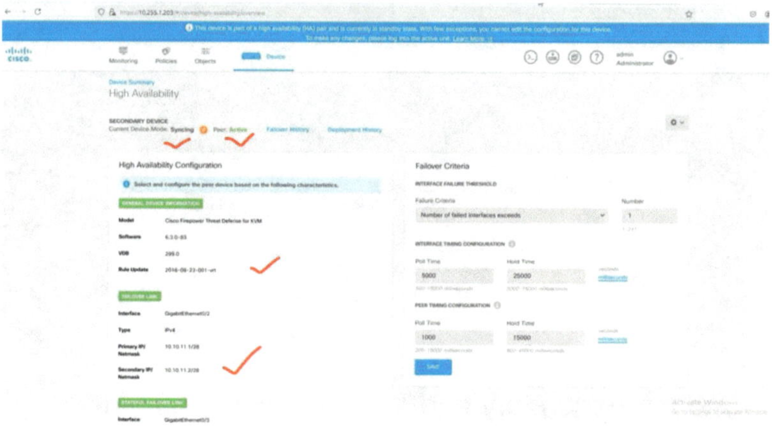

Finally is done

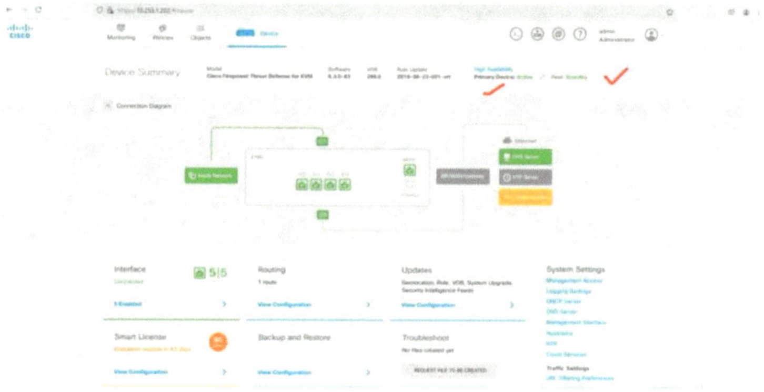

CLI verification

Show failover

Show high-availability

Author's Note

This book is the result of my **years of hands-on working and teaching experience** working in the networking and security industry. It summarizes the **most important features, configuration techniques, and real-world scenarios** that network engineers frequently encounter in production environments.

My goal has been to **distill the vast amount of Cisco knowledge** into a **concise, practical, and directly applicable guide** that will get you on track right away. Whether you are studying for Cisco certifications, configuring live systems, or strengthening your professional skills, this book serves as both a **learning companion** and a **quick-reference guide**.

You can use it **alongside official Cisco documentation** or as a **standalone resource** for mastering the most common configuration commands and concepts in Cisco networking.

For any questions about the material in this book — or to request **lab access** to practice the configurations covered — please contact me at:

Training@networkprofessional.net

Made in United States
Cleveland, OH
05 November 2026

47768321R00134